Simulations of Tropical Cyclone
in Regional Climate Models

Simulations of
Tropical Cyclone
in Regional Climate Models

Zhong Zhong
Yuan Sun

National University of Defense Technology, China

W🜚 World Scientific

NEW JERSEY · LONDON · SINGAPORE · BEIJING · SHANGHAI · HONG KONG · TAIPEI · CHENNAI · TOKYO

Published by

World Scientific Publishing Co. Pte. Ltd.

5 Toh Tuck Link, Singapore 596224

USA office: 27 Warren Street, Suite 401-402, Hackensack, NJ 07601

UK office: 57 Shelton Street, Covent Garden, London WC2H 9HE

British Library Cataloguing-in-Publication Data
A catalogue record for this book is available from the British Library.

SIMULATIONS OF TROPICAL CYCLONE IN REGIONAL CLIMATE MODELS

ISBN 978-981-3232-06-8

For any available supplementary material, please visit
http://www.worldscientific.com/worldscibooks/10.1142/10765#t=suppl

Preface

An average of about 80–90 tropical cyclones (TCs) form in tropical ocean basins worldwide each year, and roughly two-thirds of these TCs grow into typhoons or hurricanes and TCs are a critical component of tropical weather systems. The western North Pacific is the most active TC region in the world, where nearly 30 TCs form annually with more than 40% occurring in the summer. As a special component of the East Asia-western North Pacific climate system, TCs have the particularity of, at both spatial and temporal scales, being a synoptic scale weather phenomenon. Thereby, great attention has been paid to the modulation on TC activities by large-scale atmospheric circulation and ocean thermal condition, which, on the other hand, can subsequently be affected by TC activities through complicated ocean-atmosphere feedback loops. It has been proposed that the sea surface cooling induced by TCs can indirectly affect regional/global climate. While for the direct feedback on the atmosphere by TC activities, it is reflected not only in atmospheric circulation within the TC influential area, but also in energy transport, redistribution and budget following the movement of TCs, which transport huge amounts of heat and moisture from the tropics to middle and high latitudes. Moreover, effects of TC feedback on the atmosphere can spread over a much larger area via atmospheric circulation and modulate the global circulation pattern.

Uncertainties in climate simulations have been reduced by using high-resolution regional climate models (RCMs) that can better describe the topography and land surface characteristics. However, RCMs still cannot accurately simulate the mean climate state and the multi-scale climate variabilities, especially in East Asia with complex topography and large monsoon variability. For example, evaluations for the East Asian summer monsoon evolution indicate that RCMs perform well when TCs are inactive over the western North Pacific. However, capabilities of RCMs tend to deteriorate in their simulations of East Asian monsoon with increasing TC activities, which is attributed to the fact that RCMs cannot realistically capture TC tracks. In RCMs, TCs are either originated within the model domain or enter the domain through the model lateral boundaries, which makes RCM simulations of TCs quite different from the operational TC forecasts. In particular, it is impossible to provide TC circulation information by optimizing initial fields in RCM simulations. Therefore, comprehensive investigations of weaknesses in RCM simulations of TC activities and mechanism study for deviations in simulated TC track and intensity obviously have great scientific and practical values for further improvement of RCM.

This book is based on regional climate modeling studies of TC track and intensity over the western North Pacific conducted by the authors over the past decade. The contents of this book include analyses of TC track and intensity deviations simulated by RCMs, investigations of interaction between TCs and large-scale circulations, sensitivity studies of TC simulation to model physical parameterization schemes, and physical mechanism study for the deviations of TC simulations, etc.

Of course, this book will not cover all aspects of RCM in simulating TC, however, it is believed that the book provides necessary information on the issues of how the TC activities influence regional climate modeling and what is the physical mechanism the failure TC simulation exists, which is very important in interpreting the model performance over East Asia-Pacific and useful in improving model physics scheme.

This book is jointly supported by National Natural Science Foundation of China (41430426, 41605072), the R&D Special Fund for Public Welfare Industry (Meteorology) (GYHY201306025) and Jiangsu Collaborative Innovation Center for Climate Change.

Zhong Zhong and Yuan Sun
Friendship Hill, Nanjing
July 2017

Contents

List of Figures

List of Tables

Chapter 1

Effects of Tropical Cyclones on Regional Climate Modeling over East Asia in Summer

1.1 Introduction

Regional climate models (RCMs) have been increasingly used to simulate the realistic characteristics of the regional climate and to evaluate the related multiscale interactions [Giorgi and Mearns, 1991, 1999; McGregor, 1997; Wang *et al.*, 2004a]. A fundamental requirement for the development of RCMs is to examine their capability in simulating the seasonal evolution and interannual variabilities on a regional scale. This seems more important for the simulation of RCMs over East Asia due to the distinct Asian monsoon climate feature. The larger variability and uncertainty of the Asian monsoon system confirmed by extensive observational studies and numerical simulations [Ju and Slingo, 1995; Lau and Yang, 1996] suggest that the modeling research will be a challenging task because of the complex scale interactions involved.

East Asia is strongly influenced by the Asian monsoon circulation. The complexity and regional diversities of the summer precipitation over China is a universally accepted understanding in literatures. Because of the complex terrain and great variability of the monsoon climate in East Asia, simulation of the East Asian climate is sensitive to physical schemes [Hong and Choi, 2006; Zhong 2006] and the skill is still limited [Zhou *et al.*, 2008, 2009]. One of the main atmospheric systems affecting the summer precipitation over China is the western Pacific subtropical high (WPSH). The seasonal extension and withdrawal of the WPSH are responsible for the rainfall distribution over eastern China. Therefore, the simulated precipitation is directly dependent on the capability of the models to capture the activities of the WPSH and the corresponding regional circulations. In general, both RCMs and general circulation models (GCMs) can reproduce the seasonal evolution of the WPSH [Wang *et al.*, 2004b]. However, the simulated high in summertime is either weaker [Giorgi *et al.*, 1999] or stronger [Lee and Suh, 2000] than the observed one depending on the observational driven fields and the specific monsoon case. On the other hand, it is more difficult to simulate reasonably well the behavior of the WPSH over a shorter time scale during the summer monsoon period, which plays a significant role in the simulation of the precipitation over eastern China [Zhong *et al.*, 2010].

Tropical cyclones (TCs) are one of the most dangerous natural hazards to people. Every year, they cause considerable loss of life and do immense damage to property. However, TCs are essential features of the Earth's atmosphere, as they transfer heat and energy between the equator and the cooler regions nearer the poles.

In this chapter, we use the observed sea surface temperature (SST) and meteorological fields to drive a RCM to determine the degree to which the WPSH is affected by TCs and whether the RCM can reproduce the TCs and their effects properly on a regional scale. Finally, as a case study, the sensitivity experiment of the strongest TC Winnie in 1997 was conducted to illustrate its climate effect.

1.2 Effect of TC activity on regional circulation simulation

For an East Asian summertime case in 1994, the seasonal scale simulations in summer using the Regional Climate Model version 3 (RegCM3), developed by the International Centre for Theoretical Physics (ICTP), was conducted with three cumulus parameterization schemes (CPSs), i.e., the Emanuel scheme [Emanuel, 1991a], Grell scheme [Grell, 1993] and Kuo scheme [Anthes, 1977]. It was found that the model is capable of reproducing the mean summer circulation over East Asia, as has been demonstrated by previous works [Giorgi *et al.*,1999; Lee and Suh, 2000; *Liu et al.*, 1994; Kato *et al.*, 1999; Lee *et al.*, 2002; Wang *et al.*, 2003]. However, the model shows a failure simulation over shorter time scale.

Fig. 1.1 depicts the temporal evolution of the spatial abnormal correlation coefficients (ACCs) with a geopotential height of 500 hPa between the observation and simulation. The averaged ACCs in Fig. 1.1 are 0.89, 0.88 and 0.84 corresponding to different CPS, and the temporal tendency of the ACCs is basically similar in the three experiments. However, the evolution of the ACCs exhibits a sharp oscillation with three lower-value periods. This suggests that the model can reasonably reproduce the regional circulation pattern with all the three CPSs in most of the periods, but it yields some unsuccessful simulations in the cases of lower ACCs. What kind of circulation systems cannot be reproduced well by the model in the periods with lower ACCs? To gain insight into the possible cause for these lower ACCs, the selected four periods shown in Fig. 1.1, i.e., June 8–12 (period A), July 18–22 (period B), July 24–28 (period C) and August 27–31 (period D), are analyzed in detail.

The mean geopotential height of 500 hPa and the wind field for the observation and simulation with three CPSs during period B (July 18–22) are shown in Fig. 1.2. In this period, when the strong WPSH dominates over the western North Pacific (WNP) and the southwest and south monsoon flow prevail over eastern China, the mean circulation system over the eastern part of the model domain is evidently different from that in summertime. The observed circulation clearly shows that the WPSH is interrupted by a TC and the main body of the WPSH withdraws eastward. The cyclonic circulation system holds over the WNP, whereas a split anticyclonic cell from the main body of the WPSH is over eastern China and the southern Korean Peninsula (Fig. 1.2a). Obviously, the model with all the three CPSs cannot suitably reproduce the invading process of the TC and the associated split of the WPSH due to the deficient simulations of the TC activity (Figs. 1.2b–1.2d).

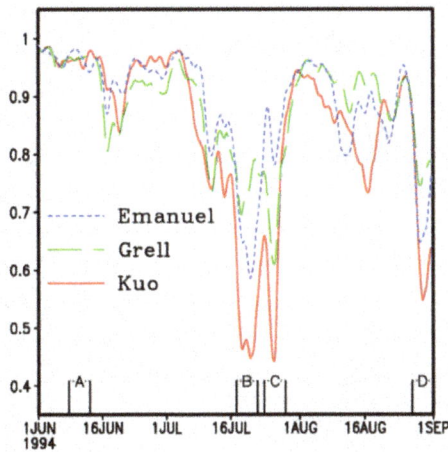

Fig. 1.1. Temporal evolution of spatial ACCs of 500 hPa geopotential height with CPSs of Emanual scheme, Grell scheme and Kuo scheme, respectively (A represents the period with higher ACC, whereas B, C and D represents the period with lower ACC, respectively).

Moreover, it can be seen that the simulated TC with the Emanuel and Grell schemes is stronger and turns northeastward in advance over the WNP to the east of Japan, whereas the simulation with the Kuo scheme gives a lower pressure system over Japan. Over the WNP where the observed TC is located, the anticyclonic circulation appears clearly. The similar situation also occurred for period B and period C in Fig. 1.1 when TC was over WNP. Therefore, for the amelioration of RCMs, the important problem of model development on how to enhance the capacity for portraying the activity of TCs in summertime over the WNP should be solved.

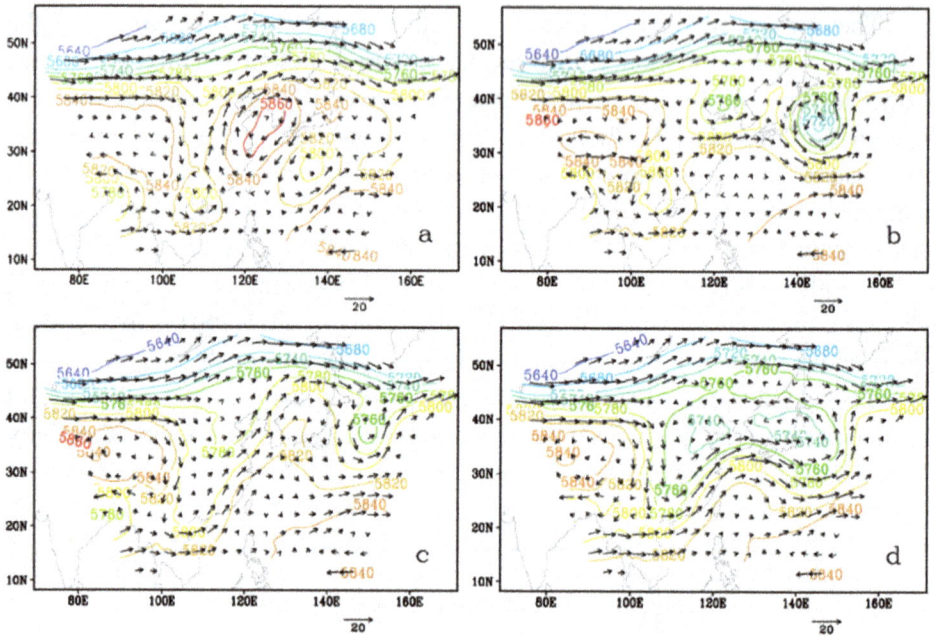

Fig. 1.2. The temporal mean geopotential height and wind vector at 500 hPa observed (a) and simulated with Emanuel scheme (b), Grell scheme (c) and Kuo scheme (d), respectively, during period B in Fig. 1.1.

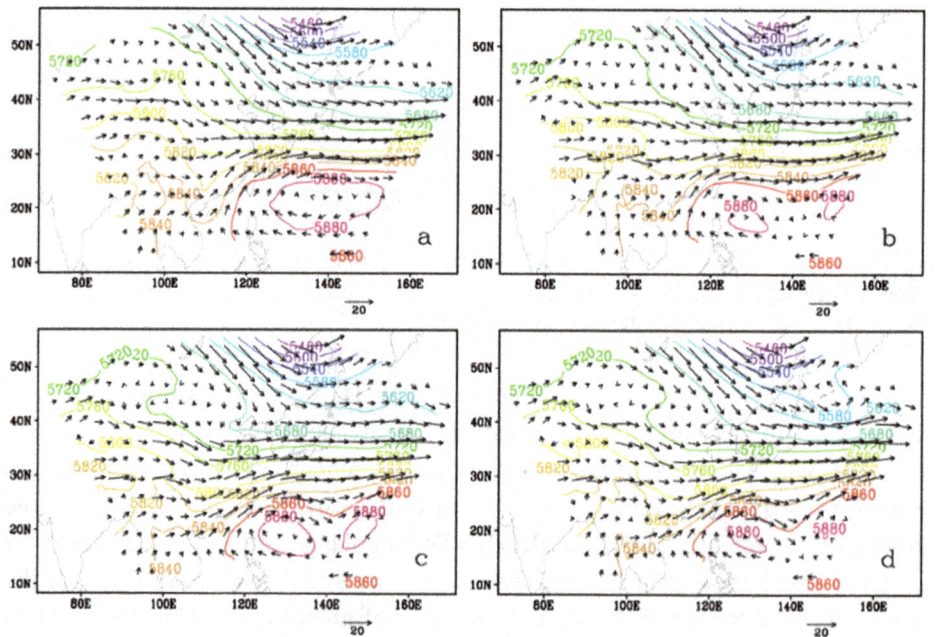

Fig. 1.3. Same as Fig. 1.2, except for the period A shown in Fig. 1.1.

On the contrary, during period A with higher ACCs as shown in Fig. 1.1, the regional circulation systems are suitably portrayed in the model because this period is in the intermittent period of the TCs in the model domain (Fig. 1.3). This clearly indicates that the simulated trough and ridge in the middle and high latitudes resemble the observed ones. The simulated WPSH in the lower latitude is also largely consistent with the observation, although the simulated one seems slightly weaker, with each of the three individual CPSs. Overall, the simulations with the Emanuel scheme or Grell scheme exhibits better results than that with Kuo scheme; the latter generates a weakened WPSH and an abnormal split trough in the middle latitude, which implies that the regional climate modeling is sensitive to model physics, such as CPS. In spite of the examples of the tropical storms as [Liu *et al.*, 1994], the model can generally reproduce their intensity and track, the regional climate modeling over East Asia in summertime on the impact of bogus typhoons implies that the model cannot suitably reproduce the impact of TCs without a special bogus technique [Lee, 2004]. TCs hinder the precise simulation of the summer monsoon circulation over East Asia [Zhong and Hu, 2007]. It is a challenge for regional climate modeling to overcome the disadvantage of the failure simulation when TCs are active over the WNP for the simulation over East Asia. In addition to the TC bogus technique, which is complicated for long-term simulation, improvements in both the treatment for lateral boundary conditions (LBCs) and model physical parameterizations could be helpful to more suitably represent the behavior of TCs in the model domain. The possible improvement may include selecting the proper buffer zone width, adopting different nudging parameters in the lower and upper troposphere, etc. With the amelioration, the TCs over the WNP would be regulated in the right manner, even though the climate model at a coarser resolution cannot describe the detailed structure of TCs.

1.3 Impact of tropical cyclone on the East Asian summer climate

It is well known that during summer, WPSH is the predominant large-scale circulation system over the WNP. In most cases, this circulation system regulates the tracks of the TCs. However, the TCs in turn influence the WPSH as well. When a stronger TC turns northeastward over the WNP, it usually causes the WPSH to split [Wu *et al.*, 2002; Zhong, 2006]. For operational prediction, the impacts of circulation systems, particularly the WPSH, on the tracks of TCs have been studied in detail. However, the quantitative effect of TCs on the regional circulation remains unknown.

Due to the limitation of available observational data, it is difficult to assess the quantitative effects of TCs on regional circulation systems, particularly on the state of regional circulation without TCs. The limited area model (LAM) is an effective tool that can be used to address this issue. It was noted that the climate in a regional model is determined by a dynamical equilibrium between two factors, i.e., the information provided by the lateral boundary condition and the internal model physics and dynamics [Giorgi *et al.*, 1999]. This suggests that the performance of the regional model is mostly dependent on the lateral boundary condition [Anthes *et al.*, 1987; Giorgi and Marinucci, 1996]. From this view point, the effects of TCs on regional circulation systems can be evaluated in terms of the removal of TCs at the lateral boundaries of the model. Therefore, the circulation systems over the model domain would be unaffected by the TCs. In this section, as a case study, RegCM3 is used to determine the extent to which the regional climate is affected by the TCs. This was done by comparing the simulation results at the climate scale with and without TCs at the lateral boundaries of the model's driven fields. The simulation begins at 0000 GMT on July 15 and ends at 1800 GMT on August 31, 1997. During the week of August 17–23 in the simulation period, China suffered huge economic losses due to the damage caused by the violent TC Winnie (1997).

Two experiments were performed, one was the control run (CTR) as described above, and the other was the sensitivity run (SER). The SER was conducted by removing the TCs from the 6h interval large-scale driven fields for the same period as that of the CTR. This strategy is in agreement with the removal technique of the large-scale TC circulation from the first-guess fields before the bogus TC is inserted into the initial fields, which is commonly used for the numerical simulation or prediction of TCs [Christopher *et al.*, 2001]. In our approach, we modified the vorticity, geostrophic vorticity and divergence. Then, we solve for the change in the nondivergent stream function, geopotential and velocity potential, and compute the modified velocity field, temperature field and the corresponding geopotential height field. The details of the strategy employed for the removal of TCs from large-scale driven fields can be found in Christopher *et al.* [2001].

Fig. 1.4 shows the monthly geopotential height and wind vector of the observations, CTR and SER at 200 hPa, 500 hPa and 850 hPa in August 1997. It is observed that the CTR performs well for the regional circulation in the lower troposphere over East Asia, while the East Asian summer monsoon predominates the southeast coast of China and the adjacent WNP; further, a cyclonic circulation interrupts the southwestern summer monsoon over south China (bottom panels in Figs. 1.4a and 1.4b). The simulated ridge line of WPSH at 500 hPa is at 30°N; this is consistent with the observations (middle panels in Figs. 1.4a and 1.4b). At 200 hPa, the mean circulation pattern and the South Asian High (SAH) are also reproduced well (top panels in Figs. 1.4a and 1.4b). However, at 850 hPa, the simulated WPSH is slightly weaker, whereas the lower depression system over the west part of northeast China is stronger than the observed one, which is partially caused by the TC Winnie passed through Bohai Sea and landed again at Yinkou, Liaoning province, and activated over northeast China.

It is noteworthy that the mean circulation in the lower troposphere in August 1997 is somewhat different from the normal summer monsoon pattern while the southwest summer monsoon flow prevails over south China [Lee and Suh, 2000]. The cyclonic circulation at 850 hPa, which interrupted the southwest summer monsoon over south China in August 1997 in the mean chart, was mainly caused by the landfall of TCs, particularly that of the violent TC Winnie on 19 August and its sweep over east China in the subsequent days.

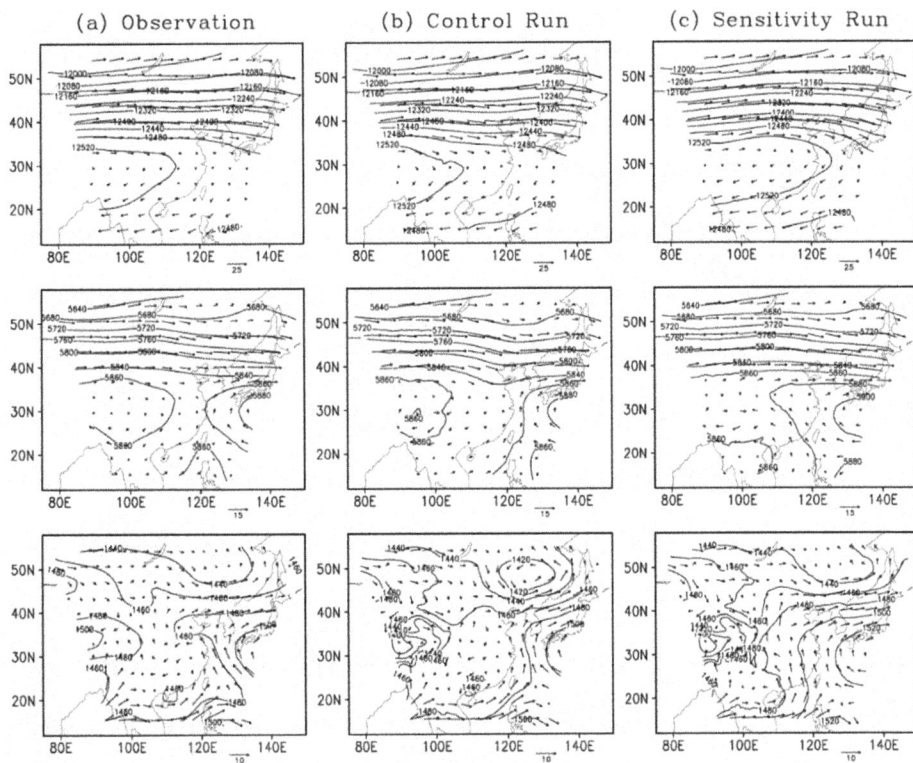

Fig. 1.4. Monthly geopotential height (contour lines) and wind (vector arrows) of the observation (a), CTR (b) and SER (c) at 200 hPa (top panel), 500 hPa (middle panel) and 850 hPa (bottom panel) in August 1997.

In the case of the SER, the simulated WPSH in the lower troposphere intensifies and extends westward significantly. Moreover, the summer monsoon in the lower troposphere with a stronger southerly component from the South China Sea (SCS) and the Bay of Bengal is predominant over the southeast mainland of China. Meanwhile, the simulated intensity of the lower depression system recovered for TC Winnie is no longer active and finally filling up over northeast China (shown in the middle and bottom panels in Figs. 1.4c). The simulated SAH at 200 hPa is also intensified. However, it extends eastward distinctly, thereby confirming that the WPSH and SAH act in the opposite directions, as noted by Wu *et al.* [2002]. The simulated mean circulation pattern for SER can be verified by the observations. As an example, Fig. 1.5 shows the monthly circulation at 200 hPa, 500 hPa and 850 hPa in August 2003, when TCs over the WNP are observed to be less than usual in this month. It also exhibits a similar pattern for the mean circulation for SER, while the SAH is intensified and extends eastward (Fig. 1.5a); further, the WPSH is also intensified, but it extends westward (Fig. 1.5b), and the southerly monsoon flow prevails over central and southern China in the lower troposphere (Fig. 1.5c).

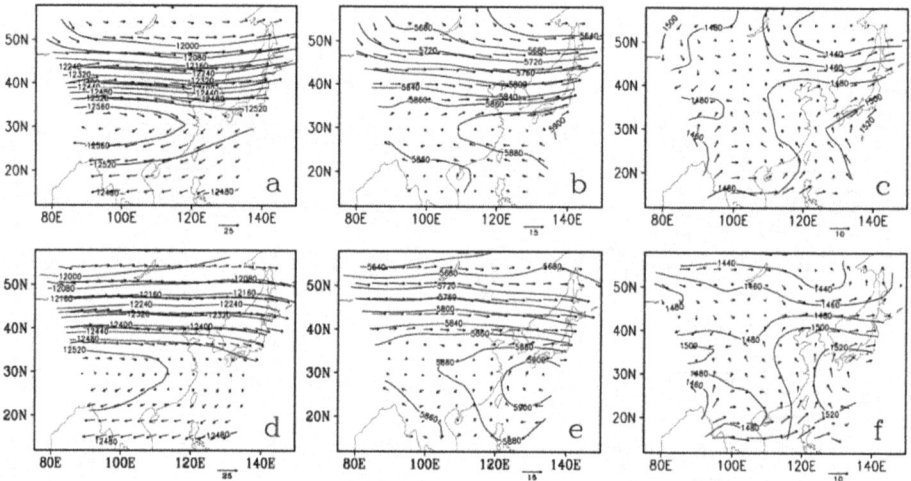

Fig. 1.5. Observed monthly geopotential height (contour lines) and wind (vector arrows) at (a) 200 hPa, (b) 500 hPa and (c) 850 hPa in August 2003, and at (d) 200 hPa, (e) 500 hPa and (f) 850 hPa in August 1997 with the observed TCs removed.

The observed monthly circulations in August 1997 with the observed TC removed run (TCR) are also shown in Fig. 1.5d–1.5f. Compared with that of SER (Fig. 1.4c), one can see that both the SAH and WPSH of TCR are stronger than that of SER, though the circulation patterns are similar with each other in the troposphere, except for the intensified high-pressure systems for TCR. Since the difference between TCR and SER is mainly on whether or not the interaction between TCs and circulation is included, it implies that the interaction itself can promote the intensification of the circulation system at middle latitude. But the interaction restrains the transportation of East Asian summer monsoon to the eastern part of China from SCS and the western Pacific Ocean (the bottom panel of Figs. 1.4c and 1.5f).

The difference in monthly circulations between SER and CTR are shown in Fig. 1.6. It clearly shows an anticyclonic circulation in the entire troposphere over eastern China and the adjacent WNP. Since TC Winnie was the most violent one in 1997, it significantly influences the difference circulation chart and it exhibits an anticyclonic difference circulation in the middle and lower troposphere and a circular anticyclonic difference circulation at 200 hPa. Moreover, the axis of anticyclonic difference circulation in the vertical is inclined westward above 500 hPa centred at approximately 120°E in the lower troposphere and over the bend of the Yellow River (40°N, 112°E) at 200 hPa.

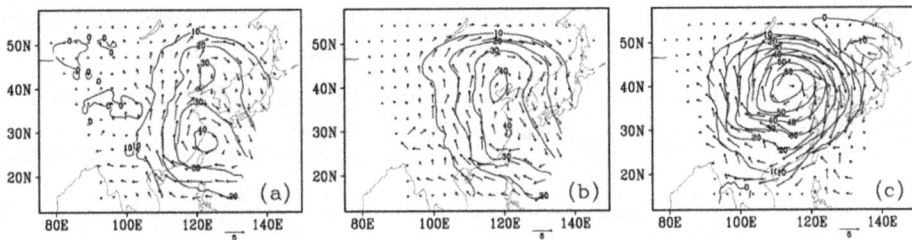

Fig. 1.6. Monthly differences in the geopotential height (contour lines) and wind (vector arrows) between SER and CTR at 850 hPa (a), 500 hPa (b) and 200 hPa (c).

In order to consider the impacts of TCs on the strength of the WPSH, here, we define the number of model grid points to the east of 110°E, where the geopotential height is greater than 5880 gpm at 500 hPa, as the intensity index of the WPSH (IWPSH). With which, the effects of the TCs on the strength of the WPSH were assessed quantitatively. Fig. 1.7 shows the temporal variations of IWPSH in the observation, CTR and SER. Four major intensification processes are observed on around August 2, 11, 22 and 30, and three weakening processes are observed on around August 7, 17 and 25. When the activities of the TCs in this period are compared, it is found that each weakening process is attributed to the presence of the following TC over the WNP: Tina, Winnie and Amber, respectively. As shown in Fig. 1.8, these TCs occurred over the WNP during three weakening process respectively and they occupied the position where the WPSH was originally located, resulting in the lower IWPSH. However, the three intensification processes (on August 2, 22 and 30, respectively) are attributed to the presence of the following TCs over the southern coastal area of China and to the west of the WPSH: Victor, Zita and Amber (as shown in Fig. 1.9a, 1.9c and 1.9d). These are beneficial to the strengthening of the WPSH over the ocean. Another intensification process (on around August 11) occurred during the intermittent period of TC over the WNP (Fig. 1.9b). This reduction in intensity when the TCs are active over the WNP demonstrates the influence of TCs on the WPSH. Here, it could be inferred that the effect of monsoon latent heating on the strength of subtropical anticyclones, as emphasized by Hoskins [1996] and Liu et al. [2001], may play a secondary role in the evolution of WPSH and it will be effective only if the TCs are not active over the WNP. Meanwhile, the latent heat release in the cloud wall around the centre of the TCs over Southeast China may have the same effect as that of monsoon latent heating, which will generate a secondary vertical circulation with an ascending branch in the cloud wall and a descending one in the WPSH. This could be the reason why the WPSH is generally intensified after a TC landfall occurs over China. A specific example is TC Amber, which weakened the WPSH when it was over the WNP, but intensified the WPSH after its landfall over China. In fact, TC Winnie also exhibited the same effect before and after its landfall; it contributed to the intensification process of the WPSH before August 21 when it was over east China.

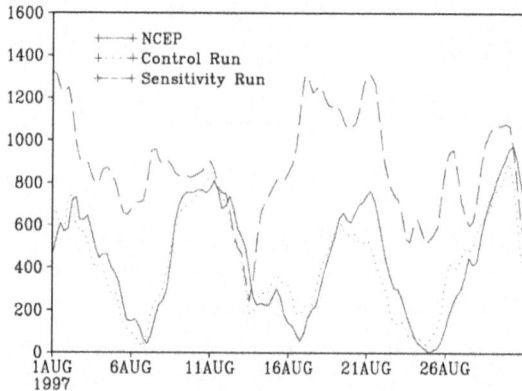

Fig. 1.7. Temporal variation in the intensity index of the WPSH for the observation (solid line), CTR (dotted line) and SER (dashed line).

The evolution of the IWPSH for CTR shows that the model can reproduce the temporal variation of the WPSH intensity, which demonstrates that the model performs well for the variation in circulation over WNP as well as for the activities of TCs. However, in SER, the simulated evolution of the IWPSH is different from that in the observed one and CTR. The most distinct feature is that the three intensification processes (on August 2, 22 and 30) are magnified except for the one on around August 11, which is not directly related to TC. On the other hand, an abnormal intensification process is observed during August 14–22 for the WPSH simulation when TC Winnie is removed from the large-scale forcing fields at the eastern lateral boundary of the model. It should be noted that the mechanisms of the observed and SER-simulated intensification processes of WPSH in those periods are different, i.e., the effects of the TCs are included in the observation but they are removed in the SER simulation. Moreover, it is revealed that the model cannot reproduce the intensification of the WPSH on around August 11, because it is not related to the activity of the TC over WNP.

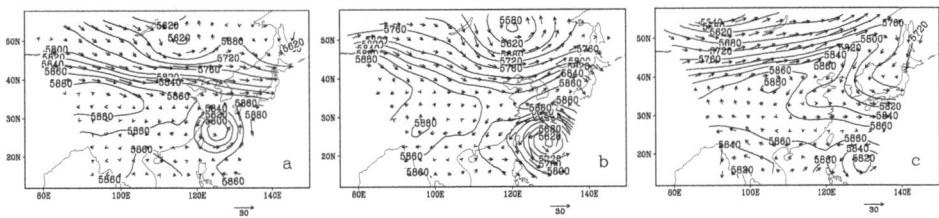

Fig. 1.8. Geopotential height (contour lines) and wind (vector arrows) at 500 hPa on August 7 (a), 17 (b) and 25 (c).

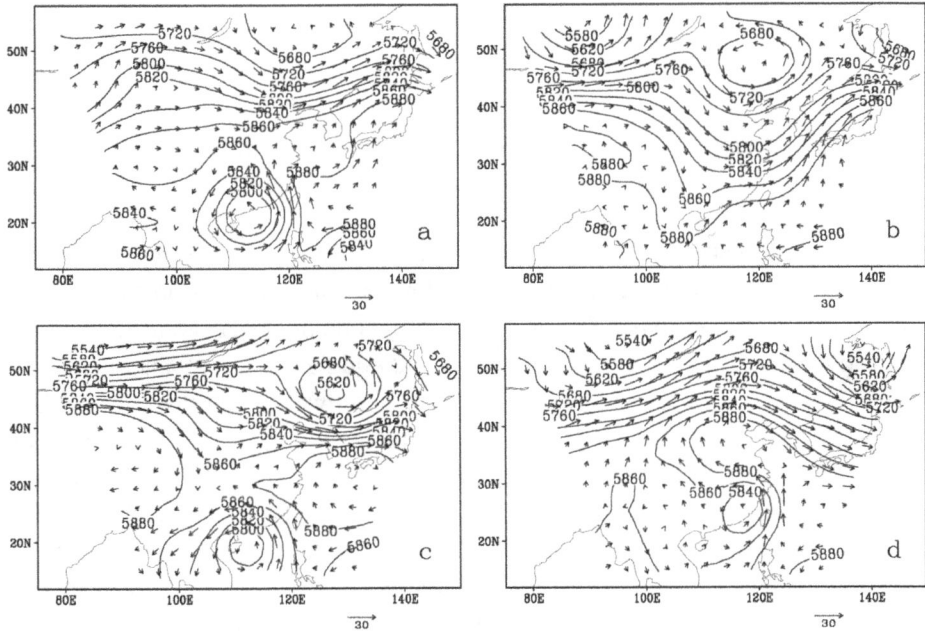

Fig. 1.9. Geopotential height (contour lines) and wind (vector arrows) at 500 hPa on August 2 (a), 11 (b), 22 (c) and 30 (d).

Generally, TCs over the WNP or China result in substantial precipitation over central and eastern China, which can mitigate the torridity and drought in East Asia when WPSH dominates over there. However, the TCs will interrupt the extension of the East Asian summer monsoon on the weather and climate scales in the lower troposphere, as shown in Fig. 1.4a and 1.4b. This will eliminate the transportation of water vapour from the Indian Ocean and SCS towards east China. Fig. 1.10 shows the monthly differences of the precipitable water vapour (PWV) and the percentage rate below 500 hPa between SER and CTR. As expected, when the TCs are removed, the atmosphere in the lower troposphere over southeast China, the adjacent WNP as well as northeast China would become drier, while that over southwest China, the eastern part of northwest China and north China would become wetter.

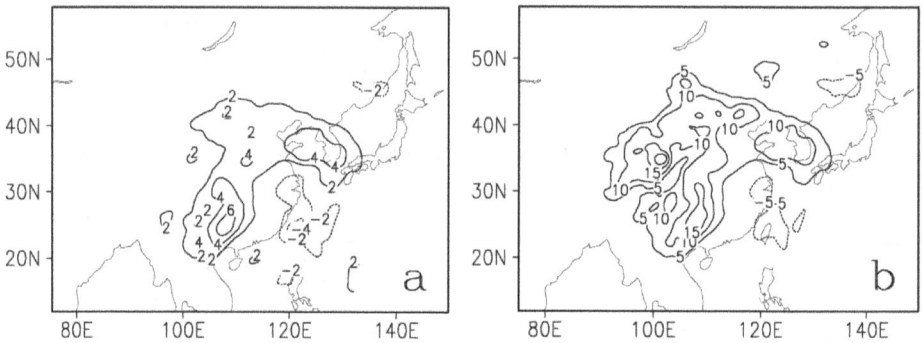

Fig. 1.10. Monthly differences in the precipitable water vapour (a, mm) and the corresponding percentage rate (b) below 500 hPa between SER and CTR.

To further illustrate the effects of TC activities on the circulation and precipitation over East Asia, the nudging technique was employed in simulations for the summer of 1998, when China was hit by a flood that occurs once every 100 years and the abnormal flooding in the lower and middle reaches of Yangtze River (LMRYR) was related to an abnormal seasonal southward withdrawal of WPSH that year. The control run and the nudging run were conducted. The nudging technique was used only in the east portion of model domain to 120°E in nudging run, which can guarantee the activity features of WPSH and TCs identical to reanalysis data as observations.

Fig. 1.11 shows the observed and simulated temporal variation of averaged ridge line of WPSH between 120°E and 130°E at 500 hPa in the summer of 1998. It can be seen that the first northward jump of ridge line was in early June, while in early July, it had been pushed to 30°N but then withdrew southward abnormally after July 10, and jumped to 30° again around the end of July, finally, withdrew southward (seasonally) in late August. The activity of WPSH was significantly abnormal in the summer of 1998, being characterized by an abnormal withdrew southward after July 10, caused "the second period of Meiyu" in the LMRYR, which lead to an unremitting flood as the rain storms poured into the Yangtze River basin when the warning water level had already been exceeded.

Although the control run could basically describe the seasonal and abnormal extension/withdrawal of the ridge line for that summer, there were significant errors in the first northward jump and in its seasonal and abnormal southward withdrawal compared with the actually activity in late June (green line in Fig. 1.11). However, in the nudging run, the extension/withdrawal of WPSH was almost identical to that observed (red line in Fig. 1.11). What should be pointed out here is that the simulated position of WPSH ridge line in the nudging run did not completely coincide with the large-scale one, which suggests that the simulated regional circulation also had a contribution to some degree, particularly in the range of 120°E–130°E, which is far from the eastern boundary of the model with a smaller nudging parameter and the simulated circulation occupying a greater proportion. Therefore, the nudging technique not only guaranteed that the extension/withdrawal of WPSH would be in accordance with the observation but also described the interaction between the WPSH and the surrounding circulation systems to some extent.

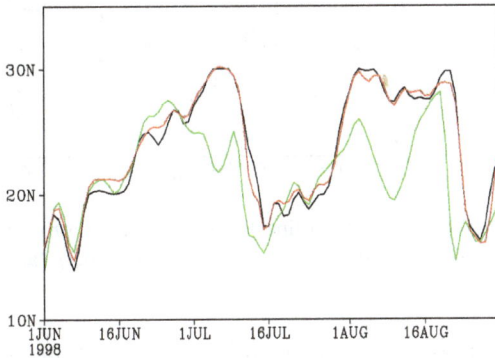

Fig. 1.11. Temporal variation of mean ridge line position at 500 hPa from observation (black), control run (green), and nudging run (red).

Here, it is further shown that the simulation error in WPSH ridge line is directly related to that of TCs. The observed and simulated daily circulations at 500 hPa on July 11, August 4, and August 10 are separately shown in Fig. 1.12, From Fig. 1.12, one can see that the positions of WPSH ridge line simulated in the control run for these 3 days are considerably different from those observed. This is because the control run could not accurately describe the activities of tropical cyclones. For example, on July 11 and August 4, Tropical Storm Nichole (1998) and Typhoon Otto (1998) were separately located over the southeastern coast of China.

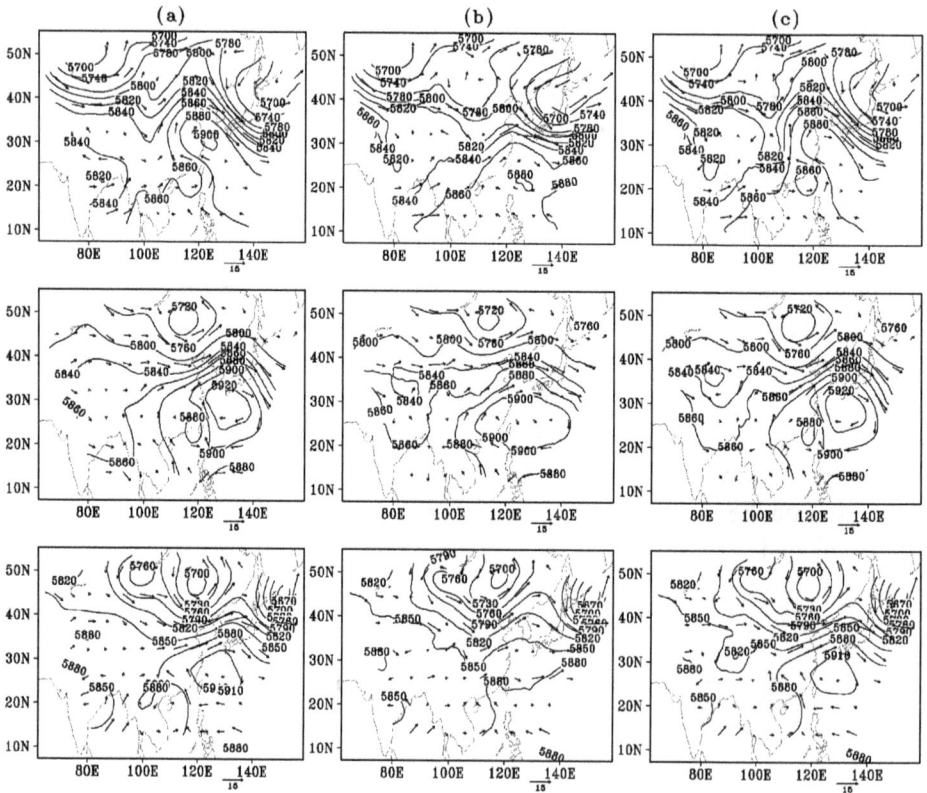

Fig. 1.12. Daily geopotential height (contour lines) and wind (vector arrows) at 500 hPa on July 11 (upper panel), August 4 (middle panel), and August 10 (bottom panel). (a: Observation, b: control run, c: nudging run).

During these periods, the control run lacked the ability to describe the TC activities, and the interaction between the TCs and WPSH could not be precisely described. Therefore, the error in the simulation of WPSH is significant and directly leads to the large error in the simulation of WPSH ridge line (upper and middle panels in Fig. 1.12). On August 10, it can be seen that there is an error in the simulation of the weaker low after the landing of Tropical Storm Penny (1998) in the control run. In fact, the error in the simulation of the whole process when Penny moved northwestward from east of the Philippines (10°N, 130°E) on August 6 and then passed through northern SCS in the control run lead to the error in the simulation of the interaction between Penny and WPSH. Therefore, the simulated intensity and position of WPSH was a departure from the truth. As for the nudging run, because the intensity and position of WPSH could be described well, the activities of the tropical cyclones could be precisely simulated (bottom panel in Fig. 1.12). Fig. 1.12 also shows that the precise simulation of the subtropical circulations allows the precise simulation of the interaction between the subtropical and mid-latitude circulations. That is, the nudging run improved the simulation of the mid-latitude circulations, especially the intensity and position of the mid-latitude low. This further confirms that the nudging technique can also partly maintain the intrinsic relationship of the interactions between surrounding circulation systems and WPSH.

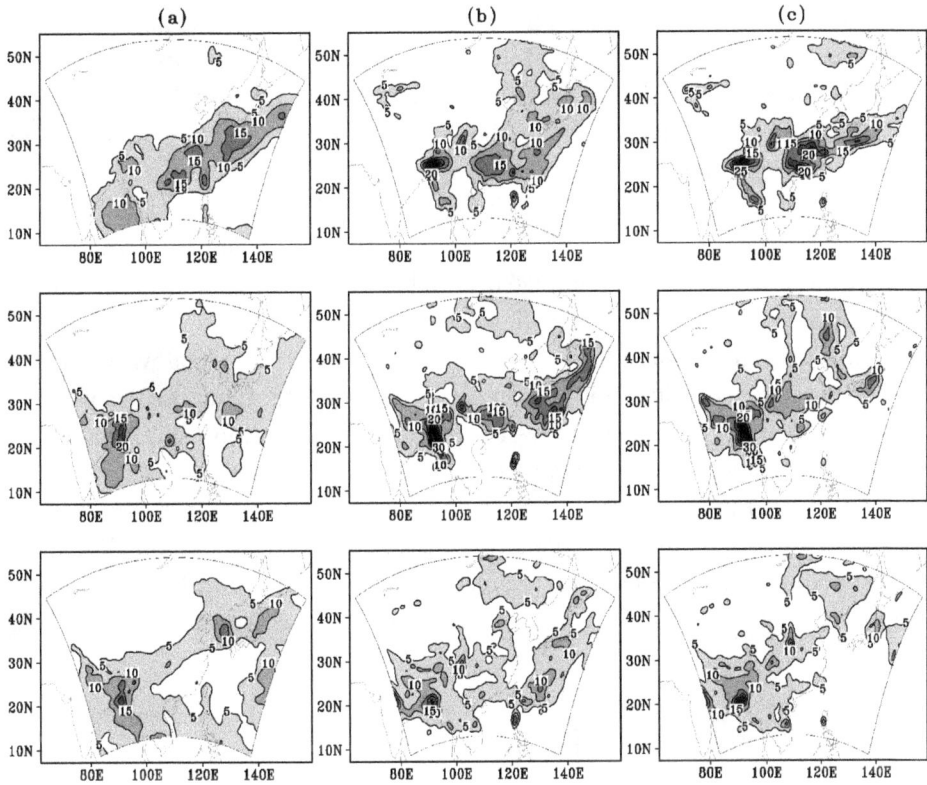

Fig. 1.13. Monthly precipitation rate (mm/day) from observation (a), control run (b), and nudging run (c) for the summer of 1998 (Upper panel: June, middle panel: July, lower panel: August).

Fig. 1.13 shows the observed and simulated monthly precipitation rates during the summer of 1998. Clearly, the nudging run reproduced the rain belt extending southwest–northeast from the Bay of Bengal to the south of Japan in

June, with three heavy rainfall areas located in southeastern China and the ocean south to Japan (upper panel in Fig. 1.13a, c). However, the control run did not reproduce the southwest–northeast extending rain belt and performed considerable error in the positions and intensities of the heavy rainfall centers, though it can capture the activity features of WPSH in June. Because the WPSH withdrew southward abnormally from July 10 to the end of July, leading to "the second period of Meiyu" in the LMRYR, an observed heavy precipitation center (more than 10mm/day) in July developed in the middle reaches of the river, while on the southeastern coast of China, the precipitation was relatively less (middle panel in Fig. 1.13a).

The heavy rainfall center in the LMRYR in July can be simulated in nudging run, except the range seemed to be larger and the heavy rainfall area extended westward to the upper reaches of the Yangtze River, where exists another heavy rainfall center in observation. Moreover, the weaker precipitation on the southeastern coast of China was precisely reproduced (middle panel in Fig. 1.13c). There was also considerable error in the simulated monthly precipitation rate in July in the control run (middle panel in Fig. 1.13b). Because the WPSH jumped northward again in August, correspondingly, the rain belt over eastern China reached the area north of the Huaihe River, leaving lesser precipitation for most of the southeastern China. The simulated monthly precipitation rate in August in nudging run is significantly better than that in the control run, especially for the rain belt from southwestern China to the lower reaches of the Yellow River and also for the weak precipitation region from southeastern China to the western Pacific. Moreover, though there were differences in position and range, the precipitation center over the Korea Peninsula and eastern Japan can be reproduced well in nudging run (lower panel in Fig. 1.13). Thus, it obviously shows that the simulation of the monthly precipitation in the nudging run for eastern China and its surrounding areas is obviously better than that in the control run, since the nudging run can capture the activities of TCs over WNP as well as that of WPSH.

Chapter 2

Lateral Boundary Buffer Zone and Its Effect on Tropical Cyclone Track

2.1 Introduction

One of the most important basic issues associated with a regional climate model (RCM) is the lateral boundary conditions [Giorgi and Mearns, 1999; Denis *et al.*, 2002; Diaconescu *et al.*, 2007]. Mathematically, the methods employed to treat an LBC in the current RCMs are ill-posed problems, and an abrupt change in grid spacing by several times at the lateral boundaries can distort the wave propagation and reflection properties [Staniforth, 1995]. Therefore, successful downscaling of RCMs from GCMs or reanalysis data requires an accurate buffer zone treatment that integrates realistic energy and mass fluxes across the RCM lateral boundaries, since the quality of the LBC has been shown to play a critical role in regional climate modeling [Giorgi and Mearns, 1999; Liang *et al.*, 2001], and minor LBC errors quickly propagate into the RCM domain and cause the model to produce unrealistic simulations [Giorgi *et al.*, 1993a; Jones *et al.*, 1995; Warner *et al.*, 1997; Gong and Wang, 2000; Diaconescu *et al.*, 2007]. In fact, the ultimate objective of the sensitivity experiment performed by Liang *et al.* [2001] for the treatment of a buffer zone was to develop high-quality LBCs.

The majority of RCMs developed to date use the so-called "relaxation" method, which was originally proposed by Davies and Turner [1977], to develop a meteorological LBC that involves the application of a Newtonian term and a diffusion term to drive the model solution toward large-scale driving fields over the lateral buffer zone. This approach is effective in ensuring a smooth transition between the LBC-dominated and the model-dominated regimes and in reducing noise generation. The size of the buffer zone can vary depending on the model domain. As pointed out by Giorgi *et al.* [1993b], the buffer zone of a few grid points width tends to produce too sharp a transition from the model solution to the driving boundary fields, and the broader buffer zone is effective in substantially reducing the noise produced by the LBC. A buffer zone of at least 10 grid points is widely used in current RCMs. Meanwhile, it has also been illustrated that large domains require broader buffer zones to ensure good consistency between large-scale circulations and driving fields [Giorgi and Mearns, 1999].

East Asia is characterized by complex topography and land-surface conditions as well as intensive human activities. The complexity of the regional climate in East Asia is induced not only by the heterogeneity of the underlying surface but also by a large-scale circulation that exhibits strong intraseasonal, interannual, and interdecadal variability [Tao and Chen, 1987]. Due to their extremely coarse spatial resolution when representing complex topography and land-surface conditions, the GCMs have serious limitations in simulating the regional climate over East Asia. For example, GCMs usually show unacceptable features for simulating the complex East Asian monsoon, especially simulating the eastern China summer rainfalls [Yu *et al.*, 2000]; none of the ten GCMs that participated in the Climate Variability and Predictability (CLIVAR)/Monsoon GCM Intercomparison Project realistically reproduced the observed Meiyu rainband [Kang *et al.*, 2002]. It is still difficult to reproduce the East Asian monsoon rainfall variability using current state-of-the-art GCMs under the specified sea surface temperature variations, and the model has limited skills in capturing the summer monsoon rainfall [Zhou *et al.*, 2008; Zhou *et al.*, 2009]. In relation to RCMs, the results of a number of studies have shown various degrees of success over East Asia by individual models [Liu *et al.*, 1997; Giorgi *et al.*, 1999; Kato *et al.*, 1999; Hong *et al.*, 1999; Wang *et al.*, 2003; Lee *et al.*, 2004; Leung *et al.*, 2004; Ding *et al.*, 2006; Qian and Leung, 2007], and the variable-resolution GCMs with local zoom over East Asia are also a helpful tools for monsoon modeling [Zhou and Li, 2002], that are useful for understanding regional climate processes and for assessing the impacts of human activities.

It is well known that large-scale wave patterns and circulation fields in an RCM are essentially determined by the driving LBC, and a relatively broader buffer zone increases the consistency between the nested model and the driving large-scale circulation patterns, both in magnitude and phase [Giorgi *et al.*, 1993b]. In the case of the East Asian summer monsoon, it was found that the RCM would provide a better simulation for various features of summer monsoon components, such as the onset of the SCS summer monsoon, the evolution of the southwest monsoon flow, and the activities of the WPSH; with a broader buffer zone, those are beneficial to the evolution of simulated summer monsoon rainfall over East Asia [Wei *et al.*, 1998]. However, previous comparison studies on the effect of different buffer zone sizes were conducted with a broader buffer zone typically configured as expanding inward from the outmost lateral boundaries of the model and keeping the model domain unchanged [Giorgi *et al.*, 1993b; Wei *et al.*, 1998; Liang *et al.*, 2001; Marbaix *et al.*, 2003]. Therefore, a broader buffer zone implies a decrease in the interior region of the model and an increase of information from the driving fields, which makes the performance of models with different buffer zones incomparable. It is difficult to determine whether or not the improvement in the simulation is only a result of the broader buffer zone size? Does the information from the driving field extending inward make a contribution to the improvement of the simulation? In this chapter, we investigate the effects of the buffer zone size on the regional climate simulation by performing a number of experiments with the buffer zone size expanding outward, while the interior region of the model for different experiments is unchanged. This would provide an objective judgment on the effect of buffer zone size. Meanwhile, the effect of lateral boundary scheme on the impact of TC cyclone track was also investigated.

2.2 Importance of lateral boundary buffer zone size on regional climate modeling

2.2.1 *Case selection*

In the summer of 1998, China suffered a natural disaster that occurs once every 100 years [Tao *et al.*, 2001], where abnormal floods in the lower and middle reaches of the Yangtze River valley (LMRYR) were directly related to the "second period of Meiyu", associated with the abnormal southward withdrawal of WPSH during the middle and late July of that year. The relationship between the evolution of WPSH and that of the rainband over eastern China in the summer has already been revealed since the 1950s, and it was found that the seasonal extension/withdrawal of the rainband over eastern China was basically synchronized with the WPSH [Tao and Chen, 1987]. Because the purpose of this study is to examine the effect of buffer zone size on the performance of RegCM3 in simulating the evolution of the East Asian summer monsoon, and the WPSH is one of the most important components of the East Asian monsoon system, the case of abnormal WPSH activity as well as the precipitation over eastern China would provide better information for estimating the capability of the model with different buffer zone sizes. Therefore, the abnormal summer monsoon case in 1998 was selected for this study.

2.2.2 *Experimental design*

Five seasonal experiments for the summer of 1998 were performed with different buffer zone sizes expanding outward. The buffer zone sizes were 5, 8, 12, 16, and 20 grid points width (represented by B05, B08, B12, B16, and B20, respectively), while the interior region of the model domain cutting out the buffer zone remained unchanged. The model domain size and the grid points for the experiments with different buffer zone sizes are illustrated in Fig. 2.1 and Table 2.1, respectively.

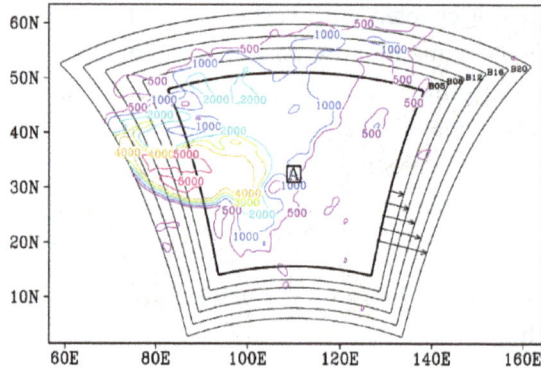

Fig. 2.1. The model domain and topography height (Units: m) for the experiments with different buffer zone size expanding outward (the thick solid line represents the inner boundaries of B05 with the buffer zone of 5 grid point width cut out).

Table 2.1. The grid points of model domain and buffer zone sizes for characteristic simulations.

Characteristic simulations	No. of grid points	Buffer zone size of grid points width
B05	80×80	5
B08	86×86	8
B12	94×94	12
B16	102×102	16
B20	110×110	20

The RegCM3 was used in this investigation. The model was driven by the meteorological boundary conditions obtained from the NCEP/NCAR reanalysis data on 2.5° latitude × 2.5° longitude at 6 h intervals [Kalnay *et al.*, 1996] and the observed monthly SSTs obtained from the UK Meteorological Office. The center of the model domain is at (32.5°N, 110°E) with a horizontal grid spacing of 60 km. The top of the model is at 50 hPa with 18 vertical levels. The seasonal simulations started at 0000 GMT on 15 May, 1998 and ended at 1800 GMT on 31 August, 1998, irrespective of the buffer zone size. The simulated regional circulations were compared with the NCEP/NCAR reanalysis data (simulations for the first 17 days were not considered to allow for the spin-up of the model). In addition, the daily precipitation from Global Precipitation Climatology Project (GPCP) data at 1° resolution [Huffman *et al.*, 2001] was used to investigate the precipitation distribution and the evolution over eastern China during the simulation period.

2.2.3 *Experimental results*

Figs. 2.2–2.4 show the seasonal mean geopotential height and wind vector in the summer of 1998 for the observations and simulations with different buffer zone sizes at 500 hPa, 850 hPa, and 200 hPa, respectively. It is observed that the model performs well for the regional circulation in the middle and lower troposphere over East Asia, while the WPSH at 500 hPa predominates the Taiwan Strait and the northern part of the SCS, and the ridge line of WPSH (defined as the line drawn through all points at which the geopotential height is maximum over the elongated high pressure belt) is several latitudinal degrees south than that in a normal year (Fig. 2.2).

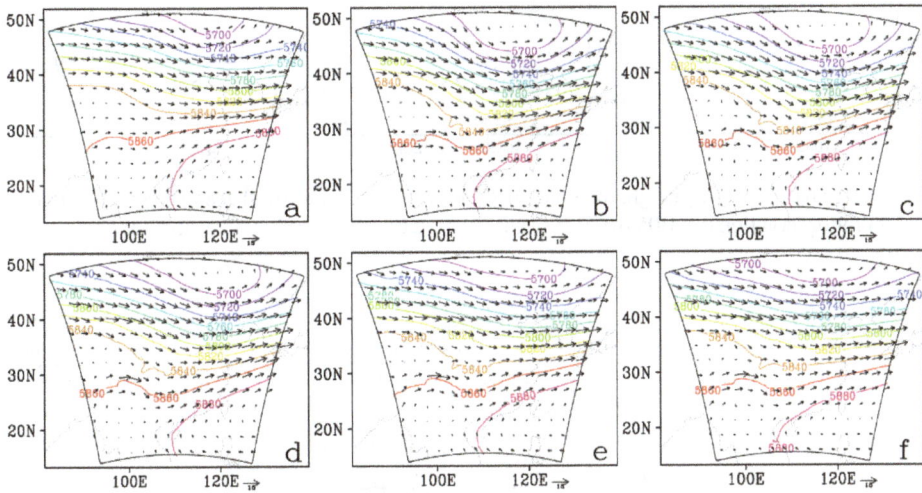

Fig. 2.2. Mean geopotential height (contour lines) and wind (vector arrows) at 500 hPa in the summer of 1998 for observation (a), B05 (b), B08 (c), B12 (d), B16 (e), and B20 (f), respectively.

At 850 hPa, the southwest monsoon flow prevailing over southern China is reproduced well; however, the simulated lower depression system over the western part of northeast China is a slightly stronger than the observed one for all experiments (Fig. 2.3). In the upper troposphere, the mean circulation pattern is also reproduced well, except for a weaker South Asian High (SAH) in all experiments performed with different buffer zone sizes (Fig. 2.4); however, the simulated intensity of SAH is increased with the outward expansion of buffer zone size.

It is difficult to estimate the differences between mean circulations with different-sized buffer zones; therefore, to examine the effects of the outward expansion of the buffer zone size, we carried out a comparison analysis for different-sized buffer zones using multiple time-scale perturbation features, the seasonal evolution of WPSH and precipitation, respectively.

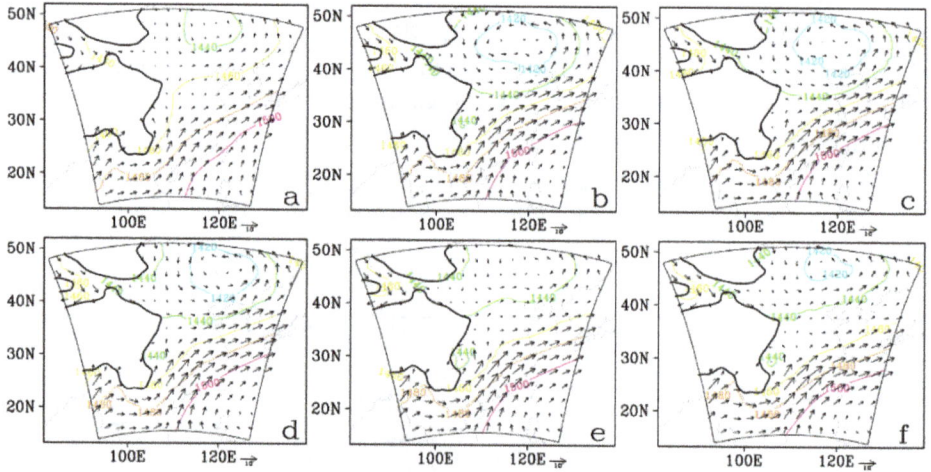

Fig. 2.3. Mean geopotential height (contour lines) and wind (vector arrows) at 850 hPa in the summer of 1998 for observation (a), B05 (b), B08 (c), B12 (d), B16 (e), and B20 (f), respectively.

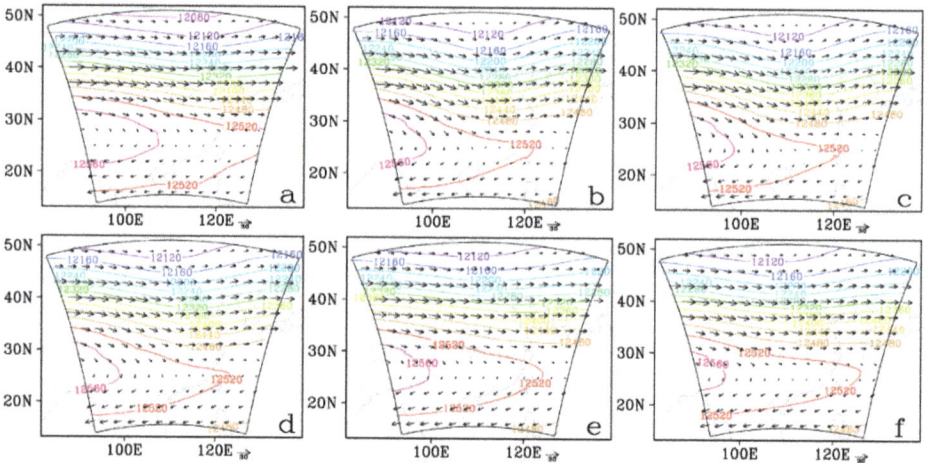

Fig. 2.4. Mean geopotential height (contour lines) and wind (vector arrows) at 200 hPa in the summer of 1998 for observation (a), B05 (b), B08 (c), B12 (d), B16 (e), and B20 (f), respectively.

2.2.3.1 *Features of multiscale perturbations*

The wavelet decomposition figures of temporal variations in area mean geopotential height at 200 hPa over a small area (square region A in Fig. 2.1) with different buffer zones are shown in Fig. 2.5. These figures were decomposed using a wavelet technique according to the perturbation periods at five time scales: 2-day, 4-day, 8-day, 16-day, and longer than 16-day (hereafter referred to as lower-frequency perturbation). They show the effect of buffer zone size on the performance of the RegCM3 for multiscale perturbations. It is evident that irrespective of the buffer zone size, the model is capable of simulating the lower-frequency perturbations; the outward expansion of the buffer zone results in an improvement in the model performance. Furthermore, the model has a certain degree of capability for simulating perturbations with periods equal to and less than 16 days (hereafter referred to as high-frequency perturbations). Despite some discrepancies between the simulated high-frequency perturbations and the observed perturbations, there are still some well-reproduced perturbations both in amplitude and in phase. However, the performance of high-frequency perturbations simulated by the model does not improve with the outward expansion of the buffer zone.

Table 2.2 lists the correlation coefficients of the time series between the simulated and the observed perturbations corresponding to Fig. 2.5. It clearly shows that with the outward expansion of the buffer zone, generally, the capability of the model for simulating lower-frequency perturbations gradually increases. The correlation coefficients of the low-frequency perturbations are greater than those of the original time series, whereas for high-frequency perturbations, no specific relationship exists between buffer zone size and the model performance. For example, the best buffer zone sizes for 2-day, 4-day, 8-day, and 16-day perturbations are 8, 16, 16, and 5 grid points width, respectively. These results suggest that a broader buffer zone size is beneficial to the large-scale (lower-frequency) circulation simulations in the upper troposphere, as proposed by Giorgi *et al.* [1993b]. In the case of high-frequency perturbations, though narrower buffer zones of 8 and 5 grid points width performed well for the 2-day and 16-day perturbations in the upper troposphere, respectively, the buffer zones of 16 or 12 grid points width generally seemed to be better for most high-frequency perturbations.

Corresponding to Table 2.2, the root of mean square errors (RMSE) is listed in Table 2.3. It can be seen that, in the upper troposphere, the RMSE of original time series is decreased with the increasing buffer zone size expanding outward when the buffer zone is less than 20 grid points width, and the RMSE for B16 is about half of that for B05. The lower-frequency perturbation gives the same feature with the original time series, and the lower the perturbation frequency, the greater the RMSE, which suggests that the simulation errors are mainly caused by lower-frequency errors of the model, although the lower-frequency variation tendency of the circulation in the upper troposphere can be reproduced well by the model.

Combined with the analysis of correlation coefficients, one can conclude that the buffer zone of 16 grid points width is better for the perturbation simulation in upper troposphere.

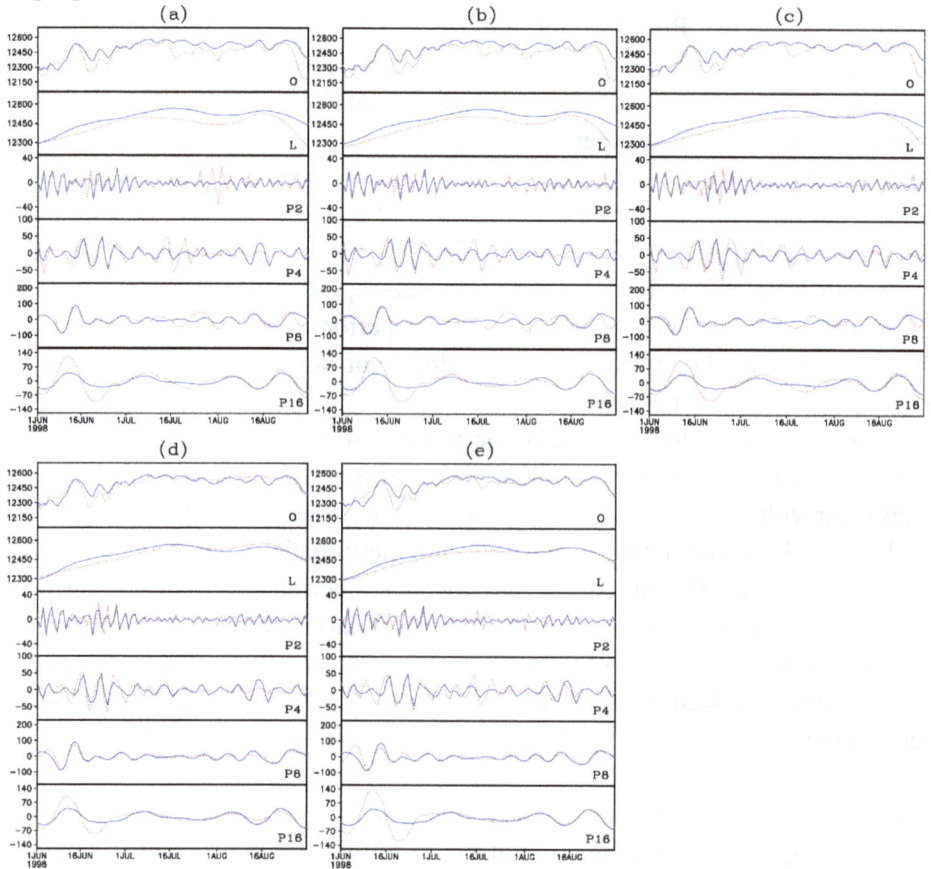

Fig. 2.5. Wavelet decomposition of observed (solid) and simulated (dotted) geopotential height averaged over area A in Fig. 1 at 200 hPa for B05(a), B08(b), B12(c), B16(d), and B20(e), where O, L, P2, P4, P8, and P16 represents original time series, low-frequency component, the component with period of 2-day, 4-day, 8-day, and 16-day, respectively.

Table 2.4 lists the correlation coefficients of the simulated and observed multiscale perturbations of area mean geopotential height at 850 hPa for region A, as shown in Fig. 2.1. It is completely different from that in the upper troposphere. Basically, the buffer zones with or fewer than 8 grid points width are better for perturbations at most timescales, except for the 4-day perturbation that had a maximum correlation coefficient with a buffer zone of 20 grid points width, but the second rank of correlation coefficient is with the 8 grid points width and is much closer to the maximum. For RMSE listed in Table 2.5, it also demonstrated that the high-frequency simulation errors could be decreased with a narrower buffer zone of 5 or 8 grid points width. Therefore, the outward expansion of the buffer zone can not improve the simulation results, and in most cases, the narrower buffer zone is better in the lower troposphere.

Table 2.2. The correlation coefficients of the original time series and the components with different periods for mean geopotential height over area A in Fig. 2.1 at 200 hPa.

	B05	B08	B12	B16	B20
O	0.7924	0.8272	0.8507	0.9003	0.8726
L	0.8970	0.9290	0.9212	0.9513	0.9708
P2	0.5613	0.6510	0.5729	0.5758	0.3901
P4	0.6769	0.6466	0.7384	0.7409	0.7126
P8	0.7170	0.7058	0.7717	0.8790	0.7135
P16	0.8784	0.8572	0.8767	0.8214	0.8001

Table 2.3. The root of mean square errors (RMSE, units: m) of the original time series and the components with different periods for mean geopotential height over area A in Fig. 2.1 at 200 hPa.

	B05	B08	B12	B16	B20
O	81.578	80.977	64.053	40.981	49.642
L	63.659	62.538	46.343	25.961	27.411
P2	9.3620	7.7411	8.4258	7.6911	8.9697
P4	16.277	17.032	15.134	13.512	14.898
P8	23.587	25.009	19.619	14.005	21.145
P16	29.565	28.356	25.716	22.438	33.961

Table 2.4. The correlation coefficients of the original time series and the components with different periods for mean geopotential height over area A in Fig. 2.1 at 850 hPa.

	B05	B08	B12	B16	B20
O	0.8515	0.8332	0.7832	0.7569	0.7894
L	0.9680	0.9683	0.9361	0.9195	0.9488
P2	0.4827	0.6012	0.3726	0.2872	0.3884
P4	0.6979	0.7275	0.7018	0.6270	0.7279
P8	0.7662	0.8426	0.7467	0.5784	0.7495
P16	0.5279	0.2245	0.2559	0.2223	0.1456

The reason may be that a broader buffer zone can draw in more lower-frequency circulation system information from lateral boundary forcing, and this capability thus restrains the development of mesoscale circulation in the lower troposphere caused by the high-resolution forcing resolved by RCMs, since the climatology of an RCM is determined by a dynamic equilibrium between the information provided by the LBC and the internal model physics and dynamics [Giorgi and Mearns, 1999].

Table 2.5. The root of mean square errors (RMSE, units: m) of the original time series and the components with different periods for mean geopotential height over area A in Fig. 2.1 at 850 hPa.

	B05	B08	B12	B16	B20
O	20.556	19.936	20.913	20.257	19.285
L	14.920	14.152	12.939	11.739	12.029
P2	5.478	5.199	6.041	5.491	5.904
P4	8.304	6.940	7.503	8.704	7.392
P8	7.382	7.338	9.654	9.184	8.180
P16	5.931	7.879	8.465	9.082	8.888

2.2.3.2 The evolution of WPSH

The temporal variations of the daily ridge line of WPSH averaged between 120° E–125° E at 500 hPa in summer for 30a (1971–2000) and 1998 are shown in Fig. 2.6. Generally, the ridge line moves northward rapidly to 20°N–25°N in middle or late June; accordingly, the rainband in eastern China also moves northward, and the LMRYR begins to enter the rainy season, i.e., the Meiyu season. In early to mid-July, the ridge line moves northward rapidly again to 25°N–30°N. At this point, the Meiyu ends, whereas the lower reaches of the Yellow River enter the rainy season. During late July or early August, the ridge line passes 30°N and reaches its northernmost position in the year. The rainband accordingly moves northward, and the northern part of northern China and northeastern China enter the rainy season. In late August or early September, the ridge line begins to withdraw, and the rainband accordingly moves southward. Throughout the summer, the ridge line shows seasonal variations by moving northward before withdrawing southward. However, it is apparent that in the summer of 1998, the ridge line of WPSH behaved abnormally, withdrawing southward in middle and late July. Accompanying such abnormal southward withdrawal, an abnormal second period of Meiyu arrived at the LMRYR associated with abnormal water vapor transport [Zhou and Yu, 2005], where the river had already reached warning levels, and the excess heavy rainfall resulted in flooding that occurs once every 100 years [Tao *et al.*, 2001].

Fig. 2.7 shows that the observed and simulated temporal variations in the ridge line of WPSH averaged between 120°E–125°E at 500 hPa in the summer of 1998. It can be seen that although the experiments with different buffer zone sizes could depict the abnormal southward withdrawal of the ridge line from mid to late

July, the deviations in the simulated ridge line become increasingly significant as the buffer zone expands outward to a width of greater than a 5 grid points width. The correlation coefficients between the simulated and the observed temporal evolution of the ridge lines are 0.8692, 0.9089, 0.8352, 0.8381, and 0.7422 for experiments B05, B08, B12, B16 and B20, respectively. Moreover, when the buffer zone is 16 or 20 grid points width, the movement of the simulated ridge line from early July showed a significant deviation from that of the observation The simulated features of WPSH at 850 hPa are almost the same as those at 500 hPa (figures not shown). It can be concluded that an outward expanding broader buffer zone is not beneficial to the activities of WPSH in the middle and lower troposphere.

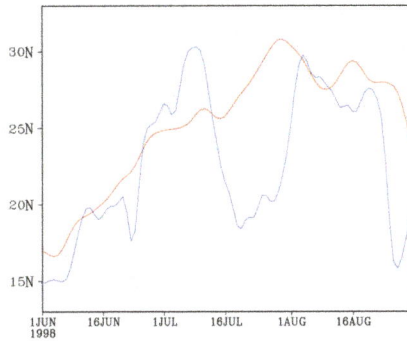

Fig. 2.6. Temporal variation of the ridge line of WPSH averaged between 120°E-125°E at 500 hPa for the summer of multi-year mean (red) and 1998 (blue).

Therefore, from the viewpoint of the multitemporal scale perturbation simulation, a broader buffer zone is only beneficial to the low-frequency perturbations in the upper troposphere; however, in relation to circulations in the lower and middle troposphere, the buffer zone should not be very wide. This also verified the configuration introduced by Giorgi *et al.* [1993b], in that a broader buffer zone in the upper troposphere and a narrower one in the middle and lower troposphere are reasonable, despite the fact that the issue of comparability among the experiments exists with inward increments in buffer zone size.

2.2.3.3 Precipitation

Fig. 2.8 gives the observed and simulated distribution of the seasonal mean precipitation rate in the domain D05, the interior region of B05 with the buffer zone of 5 grid points width cut out. It is apparent that the model cannot reproduce intensive precipitation greater than 10 mm/d in LMRYR, caused by the second period of Meiyu associated with the abnormal southward withdrawal of WPSH, and the model also demonstrates its weakness in simulating the precipitation over southern China. With the outward expansion of buffer zone size, the correlation coefficients between the simulated and the observed seasonal mean precipitation pattern in D05 are 0.4879, 0.5530, 0.6149, 0.6343, and 0.6378.

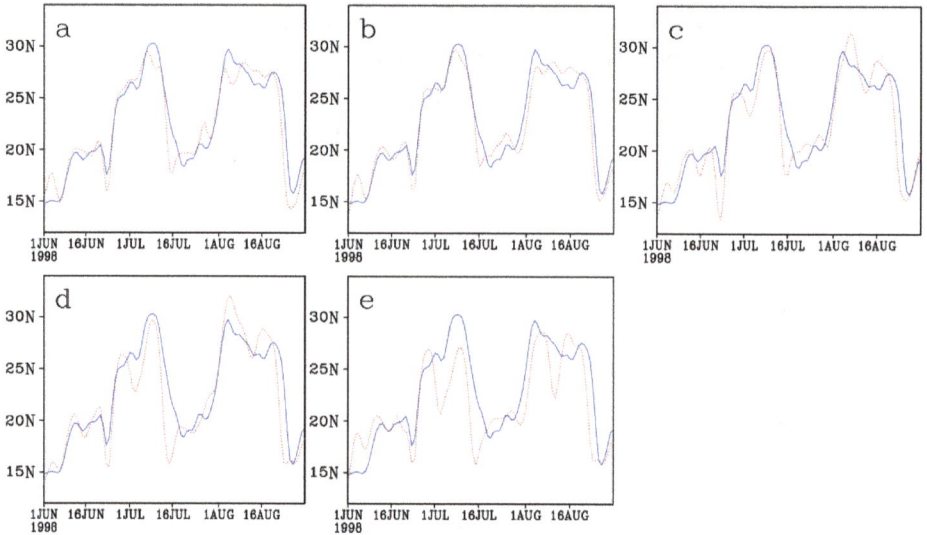

Fig. 2.7. Observed (solid) and simulated (dotted) temporal variation of the ridge line of WPSH averaged between 120°E-125°E at 500 hPa for the summer of 1998 for B05(a), B08(b), B12(c), B16(d), and B20(e), respectively.

The model shows better performance in precipitation simulations when a broader buffer zone is used. However, for the correlation coefficient, the model's performance for precipitation simulation differs with the calculation domain. For the domains D10 and D15 shown in Fig. 2.8, the interior regions of B05 with the outmost 10 and 15 grid points width cut out, respectively, the correlation coefficient over domain D10 (D15) becomes 0.5582 (0.6144), 0.5961 (0.6220), 0.5966 (0.6053), 0.5942 (0.6091), and 0.5640 (0.5640) for B05, B08, B12, B16, and B20, respectively. This shows that the degree of correlation coefficient of precipitation simulation over the central part of the model domain does not increase with expansion in the buffer zone, and a broader buffer zone is only favourable for the precipitation simulation in regions near the buffer zone. This is particularly apparent from the reduction in spurious precipitation shown in the northeastern boundary of Fig. 2.8b with a narrower buffer zone size, which suggests that the edge effects on the precipitation simulation can be constrained with a broader buffer zone. Combining the wavelet decomposition results with the results of circulation in the lower troposphere, it can be concluded that a broader buffer zone does not affect the performance of the model in the middle and lower troposphere, whereas the model's physics and dynamics that are associated with complex topography and land use may play a more important role in the circulation and precipitation simulation.

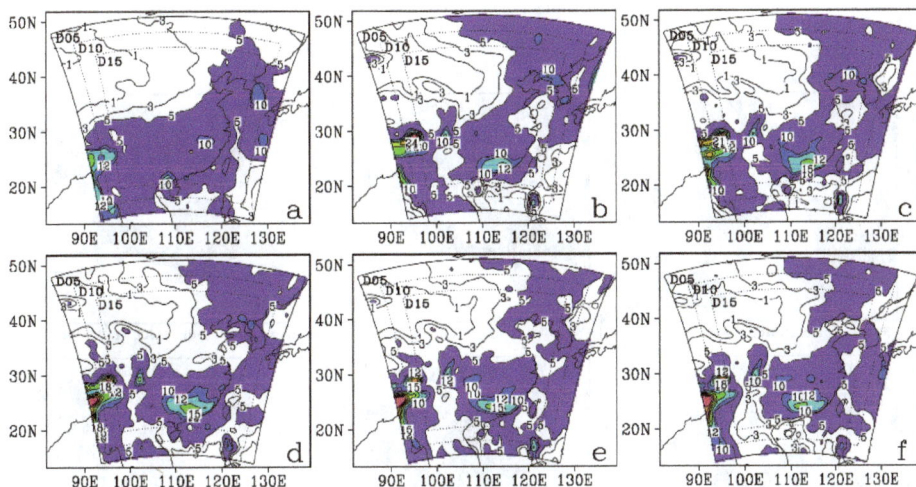

Fig. 2.8. Observed and simulated seasonal mean precipitation rate (mm/d) for domain D05 (where D05, D10, and D15 represents the interior region of B05 with the 5, 10 and 15 outmost grid points width cut out, respectively). (a: GPCP; b: B05; c: B08; d: B12; e: B16; f: B20).

To show the effect of buffer zone size on the seasonal extension/withdrawal of the rainband over eastern China, we depicted the time-latitude section of the precipitation rate greater than 5 mm/d averaged between 110°E–120°E (Fig. 2.9). The observed extension/withdrawal of the rainband shows that it gradually moves northward from early June through the first ten days of July. An abnormal southward fallback then occurred, and this persisted for about 15 days and resulted in "the second period of Meiyu" until it pushed northward again in the last ten days of July. The rainband completed is seasonal southward withdrawal in the last ten days of August (Fig. 2.9a). Comparing Fig. 2.9b–9f to Fig. 2.9a, one can see that the evolution of the rainband over eastern China cannot be reproduced well by the model, especially the abnormal southward withdrawal during the ten days in mid-July. The correlation coefficients for the observed and simulated precipitation rates on the time-latitude sections are 0.3162, 0.3214, 0.3342, 0.3285 and 0.3639 for B05, B08, B12, B16, and B20, respectively. It seems that the model performs even better at some degrees for the evolution of the rainband as the buffer zone expanding outward basically, which differs from that of the ridge line of WPSH.

Therefore, the key difference apparent from the observed relationship between the evolution of the rainband over eastern China and WPSH, is that the performance of the model in simulating the evolution of the rainband over eastern China is not completely dominated by the WPSH. This also implies that the precipitation physics of the model is more important than the dominant circulation of WPSH for reproducing the precipitation processes during the evolution of the rainband over eastern China.

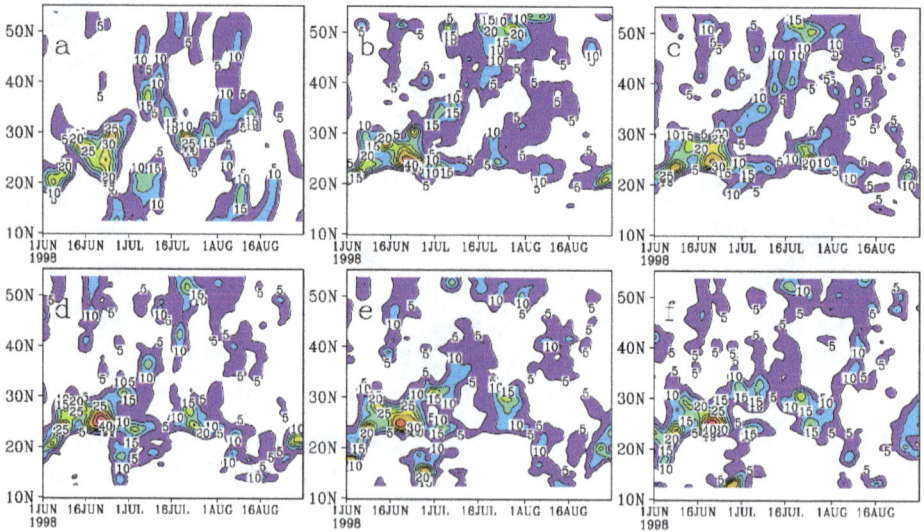

Fig. 2.9. Observed and simulated time-latitude section of daily precipitation rate greater than 5mm/d averaged between 110°E-120°E (a: GPCP; b: B05; c: B08; d: B12; e: B16; f: B20).

2.2.4 *Influences of the large-scale driving field*

The lateral boundary treatment decisively governs the possible solutions that can be generated by the model [Jones *et al.*, 1995; Giorgi *et al.*, 1999; Liang *et al.*, 2001; Marbaix *et al.*, 2003]. In this section, for the case of the abnormal flooding event in China that occurred in the summer of 1998, a comparison of the ability of different buffer zone sizes to simulate the event was investigated by performing five regional climate modeling experiments with buffer zone sizes expanding outward. The results were objectively evaluated by considering the effects of the buffer zone size. The comparison analysis shows that a broader buffer zone is only favourable to low-frequency (large-scale) circulation systems in the upper troposphere, and it is not effective in reproducing circulations in the middle and lower troposphere and reproducing the precipitation distribution. The broader the buffer zone, the greater the discrepancy in the extension/withdrawal of the ridge line of WPSH, and the lower the correlation coefficient of the precipitation pattern between simulation and observation. It suggests that the reason why the broader buffer zone is in favour of the model's performance by previous studies, may be in virtue of much large-scale information. It also demonstrates that the model's performance in relation to the seasonal evolution of the rainband over eastern China is not completely consistent with that of WPSH, which implies that precipitation physics seems more important for reproducing the details of precipitation than those of dominant circulation systems.

Note that as indicated by Diaconescu *et al.* [2007], when the large-scale driving data includes errors, the mesoscale circulation systems developed in the model domain also contain errors. As the buffer zone expands outward, more large-scale driving data errors may be introduced; this will distort the result obtained from the experiments. Moreover, since the performance of the RCMs is related to the domain size as well as to the lateral boundary data errors, it is necessary to carry out simulations for different domain sizes provided by the "perfect" LBCs.

In summer, it has been shown that the activity of TCs over the ocean is very important for the formation of the global/regional climate [Sriver *et al.*, 2007; Zhong and Hu, 2007]. A possible cause of the RCM's failure in simulating the eastern Asian summer monsoon is the fact that the tracks of TCs over the western North Pacific cannot be captured well by the model [Zhong, 2006], whereas the precipitation induced by TCs has a significant impact on eastern China [Hsu *et al.*, 2008]. Because the inward expansion of the buffer zone can introduce the "perfect" large-scale information, such as the tracks of TCs, into the model domain, it is not clear whether a broader buffer zone itself helps in improving the simulation of the summer monsoon as well as the precipitation evolution over eastern China, or the large-scale information from the lateral boundary contributes significantly to the simulation. Results of the experiment performed using a buffer zone that expands outward objectively indicates that despite the broader buffer zone guaranteeing the model's performance for large-scale circulation in the upper troposphere, it could not enhance the capability of the RCM in simulating the summer monsoon in the middle and lower troposphere as well as the precipitation distribution and its evolution over eastern China.

2.3 Effect of lateral boundary scheme on the TC track simulation

Generally, a good performance of the RCM is dependent to the good LBCs. The success of the RCM depends on the adoption of appropriate lateral boundary technique [Giorgi and Mearns, 1999; Denis *et al.*, 2002; Diaconescu *et al.*, 2007]. Previous studies have shown that it is more important to provide a good lateral boundary condition for the RCM than to improve the physical process schemes. The adoption of a broader buffer zone size (BZS) has been recommended for the simulation of the East Asian summer monsoon using the RCM [Wei *et al.*, 1998]. Majority of RCMs developed to date use the so-called nudging technique, which was originally proposed by Davies and Turner [1977], to develop a meteorological LBC that involves the application of a Newtonian term and a Laplace diffusion term to drive the model solution toward large-scale driving fields over the lateral boundary buffer zones, in insuring a smooth transition between LBC-dominated and model-dominated regimes and in reducing noise generation [Giorgi, *et al.*, 1993b]. As the climate of the regional model is the equilibrium of the atmospheric physical and dynamical processes and the information provided by the LBCs [Giorgi *et al.*, 1999], then whether or not the lateral boundary scheme affects the track of the TC in the model domain is a technical problem of the RCM.

Tropical cyclone Winnie (1997), which developed over the central Pacific on August 8, 1997, was the strongest TC in the Western Pacific in that year. It brought considerable loss of life and property and adversely affected China's national economy because Winnie passed through eastern China after its landfall at Wenling, Zhejiang Province, on August 18. Taking August of 1997 as an example and using RegCM3, we will examine the impact of lateral boundary buffer zone scheme on the ability of the RCM to describe TC activity in an effort to provide a basis for improvement of the simulation of the East Asian summer monsoon climate.

2.3.1 *Experimental design*

The dynamic core of RegCM3 is equivalent to the hydrostatic version of the NCAR/Pennsylvania State University mesoscale model, MM5 [Grell *et al.*, 1994]. The physical parameterizations employed in this simulation include the comprehensive radiative transfer package of the NCAR Community Climate Model CCM3 [Kiehl *et al.*, 1996], the nonlocal boundary layer scheme of Holtslag *et al.* [1990], the BATS land surface scheme [Dickinson *et al.*, 1993], and the cumulus parameterization scheme of Grell [1993] with the Fritsch-Chappell-type closures. The model domain is centred at (32.5°N, 120°E) with 121 east-west points and 80 north-south points and a horizontal grid spacing of 60 km. The top of the model is at 50 hPa with 18 uneven levels in the vertical.

A set of simulation experiments were performed to addresses the nudging parameters of the model prognostic equations within the buffer zone. RegCM3 adopts nudging technology to introduce the large-scale forcing in the buffer zone of lateral boundaries. For a variable T, the nudging equation in the buffer zone can be written as

$$\frac{\partial T_M}{\partial t}(n) = F(n)F_1(T_L - T_M) - F(n)F_2\nabla^2(T_L - T_M) \tag{2.1}$$

In equation (2.1), subscripts L and M refer to the large-scale forcing field and model simulation field, respectively. $F(n)$ is a function of the buffer zone ordinal n inward from lateral boundaries. $F_1 = a/\Delta t$ and $F_2 = (\Delta S)^2/(b\Delta t)$, where nudging parameters a and b has a default value of 0.1 and 50, respectively, can be adjusted to make a strong nudging (approaching to forcing fields rapidly) or weak nudging (approaching to forcing fields slowly). Δt and Δs are the time step and horizontal grid spacing of the model, respectively. With the default value of nudging parameters a and b, the experiments for different buffer zone size suggest that the best BZS is 14 for case of Winnie (1997). Using the best BZS, the experiments for the effect of nudging parameters in equation (2.1) set a is 0.05, 0.075, 0.1, 0.15 and 0.2 and b is 100, 66.7, 50, 33.3 and 25, respectively. Hence, there are a total of 25 experiments with different configurations of nudging parameters a and b. Table 2.6 lists the experimental names and the corresponding configurations of nudging parameters a and b.

Table 2.6. Experimental names and the configuration of nudging parameters a and b.

name	NP01	NP02	NP03	NP04	NP05
a/b	0.05/100	0.05/66.7	0.05/50	0.05/33.3	0.05/25
name	NP06	NP07	NP08	NP09	NP10
a/b	0.075/100	0.075/66.7	0.075/50	0.075/33.3	0.075/25
name	NP11	NP12	NP13	NP14	NP15
a/b	0.10/100	0.10/66.7	0.10/50	0.10/33.3	0.10/25
name	NP16	NP17	NP18	NP19	NP20
a/b	0.15/100	0.15/66.7	0.15/50	0.15/33.3	0.15/25
name	NP21	NP22	NP23	NP24	NP25
a/b	0.20/100	0.20/66.7	0.20/50	0.20/33.3	0.20/25

RegCM3 was employed to conduct the 1 month long simulation for August 1997. The experiments examine the effect of adjusted nudging parameters a and b in the buffer zone on TC track simulation. Each experiment starts at 00:00 GMT on August 1 and ends at 18:00 GMT on August 31. The initial and LBCs were provided by National Centres for Environmental Prediction/National Centre for Atmospheric Research (NCEP/NCAR) reanalysis data, and the LBCs were updated at 6 h interval. Sea surface temperature data are taken from the Global sea-Ice and Sea Surface Temperature (GISST) of the Hadley Center and updated once a week. In addition, the daily precipitation from Global Precipitation Climatology Project (GPCP) data at 1° resolution was used to evaluate the precipitation distribution during the simulation period. The integration time step is 180 s. We specifically analyse the simulation result for Winnie in the model region from August 12 to August 20, whereas the simulations for the first 11 days are not considered to allow for the spin-up of the model [Zhong *et al.*, 2007].

2.3.2 *Effects of nudging parameter*

In this section, we will investigate the effect of nudging parameters a and b on the performance of the model to the track simulation of Winnie, as well as the temporal variation of WPSH, with the best buffer zone of 14 grid points width. A total of 25 experiments were conducted for different configuration of nudging parameters a and b as listed in Table 2.6.

2.3.2.1 *The track of Winnie*

Fig. 2.10 presents the simulated Winnie tracks for 25 experiments with different configuration of the nudging parameters a and b. It clearly shows that the nudging parameter has a great impact on the simulation of Winnie track, and an appropriate configuration can effectively improve the track simulation. Among all experiments, NP19 ($a = 0.15$ and $b = 33.3$) shows its well performance for the westward track of Winnie entering the East China Sea, though the track error is still significant. However, an inappropriate configuration results in a much greater error in simulating the track; accompanying with a more obvious turning ahead of time and landfall on southern Japan. The RMSEs of 25 experiments are listed in Table 2.7, which also presents that NP19 performed best among all experiments with different nudging parameter configurations, and its RMSE is less than half of the NP05. Therefore, although the problem of the turning ahead of time in the simulation cannot be solved fundamentally by configuring the nudging parameters, an appropriate configuration can effectively reduce the simulation error of TC track. On the other hand, it can also be seen from Table 2.7 that the ratio of the two nudging parameters a and b is an important reference in the configuration of nudging parameter. It suggests that if a smaller value is selected for parameter a, then a larger value must be selected for parameter b, and vice versa, which implies that, within the buffer zone, two additional terms with equilibrium relationship in nudging equation (2.1) are necessary for the better performance of the model. In contrast, the track simulation would be suboptimal. Furthermore, the default setting of the two nudging parameters of the model keeps the equilibrium relationship. When the equilibrium relationship is achieved, either strong nudging experiment (a is too large and b too small) or weak nudging experiment (a is too small and b too large) is beneficial for the simulation of the Winnie track. When a or b exceeds the value range in Table 2.7, the model integration would be overflow, which is in accordance with that proposed by Marbaix *et al.* [2003].

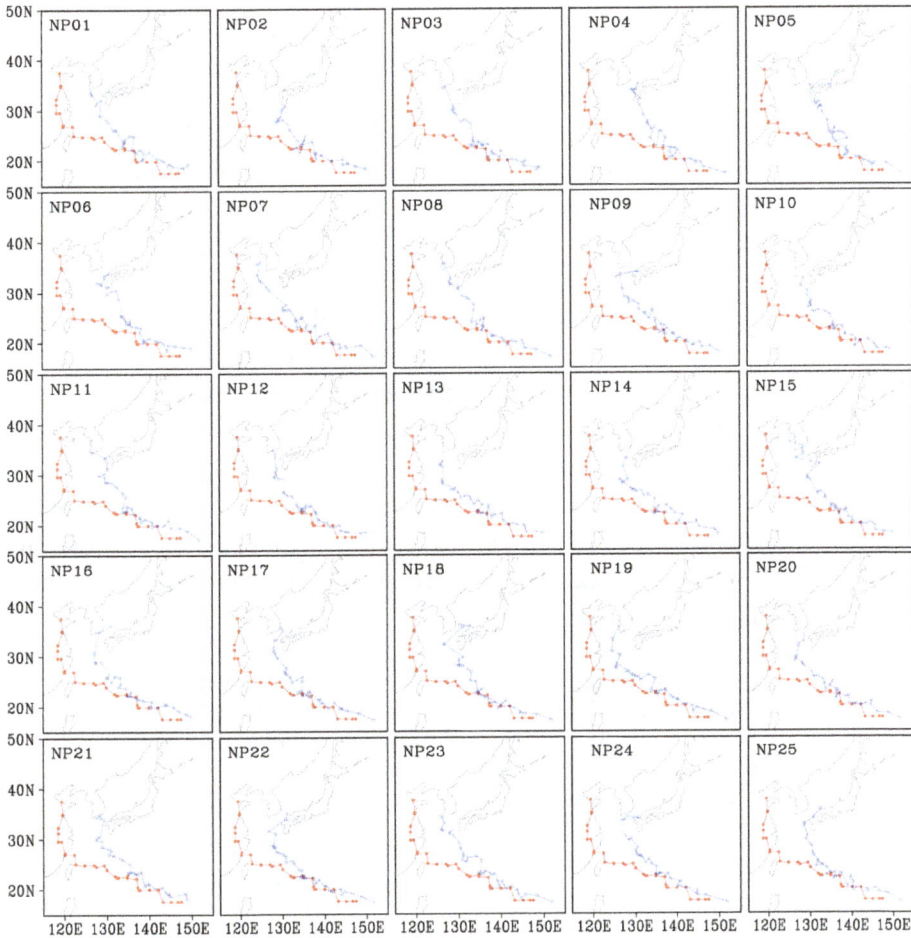

Fig. 2.10. Observed (filled circle) and simulated (open circle) tracks of Winnie at 6 h interval from 00:00 GMT on August 12 to 12:00 GMT on August 20, 1997, for 25 experiments with different configurations of nudging parameters as listed in Table 2.6.

2.3.2.2 The intensity of WPSH

Fig. 2.11 shows the observed and simulated temporal variation of the intensity index of WPSH for the 25 experiments configuring with different nudging parameters. It is shown that although the simulated evolution of the intensity index differs for different parameter configuration in detail, the basic feature of the evolution for each experiment is not substantially different. The correlation coefficients for the time series of intensity index of WPSH between observation and simulation experiment from NP01 to NP25 is 0.8168, 0.7964, 0.8495, 0.7813, 0.8038, 0.8098, 0.8504, 0.8365, 0.8270, 0.8455, 0.8106, 0.8411, 0.7873, 0.8227, 0.8615, 0.7975, 0.8331, 0.8554, 0.9027, 0.8239, 0.8382, 0.8350, 0.8957, 0.8641 and 0.8080, respectively. Here again, the NP19 obtains the best performance for the temporal variation of the intensity index of WPSH, meanwhile, the simulated intensity index shows its variation in well agreement with the observed one before 12:00 GMT of August 14, and the weak degree of the simulated WPSH is smallest after August 17 among all 25 experiments.

Furthermore, same as the experiments for optimal BZS, the sharply variation of the intensity index on August 17 can not be captured also for all 25 experiments with different configuration of nudging parameters, and the simulated intensity index transition appears on August 16. In addition, all the experiments show their higher intensity index before 12:00 GMT on August 14, and lower intensity index after 00:00 GMT on August 15 than that of the observation. Therefore, it could be concluded that no experiment with different configuration of nudging parameters could solve the problem of turning of the Winnie ahead of time fundamentally. But an appropriate configuration can be effective insuring the simulated track more being close to the observed one.

Table 2.7. RMSEs between the observed and simulated track of Winnie for the experiment with different configuration of nudging parameter a and b. The underlines highlight minimum RMSEs for the configuration of nudging parameters (units: km).

a \ b	25	33.3	50	66.7	100
0.05	512.6	436.1	367.8	409.0	334.9
0.075	377.6	327.9	365.8	<u>276.8</u>	412.3
0.10	357.1	309.8	<u>297.6</u>	332.8	374.7
0.15	308.6	<u>248.6</u>	418.9	366.2	346.1
0.20	389.0	360.3	303.6	353.6	339.8

From the simulated pattern of the precipitation rate averaged between August 12 to August 19, one can see that although an appropriate nudging parameter configuration could partly improve the precipitation pattern simulation, it does not make the simulation being consistency with the observation entirely, which implies that the description of the precipitation physics in the model is still vital in simulating the precipitation pattern.

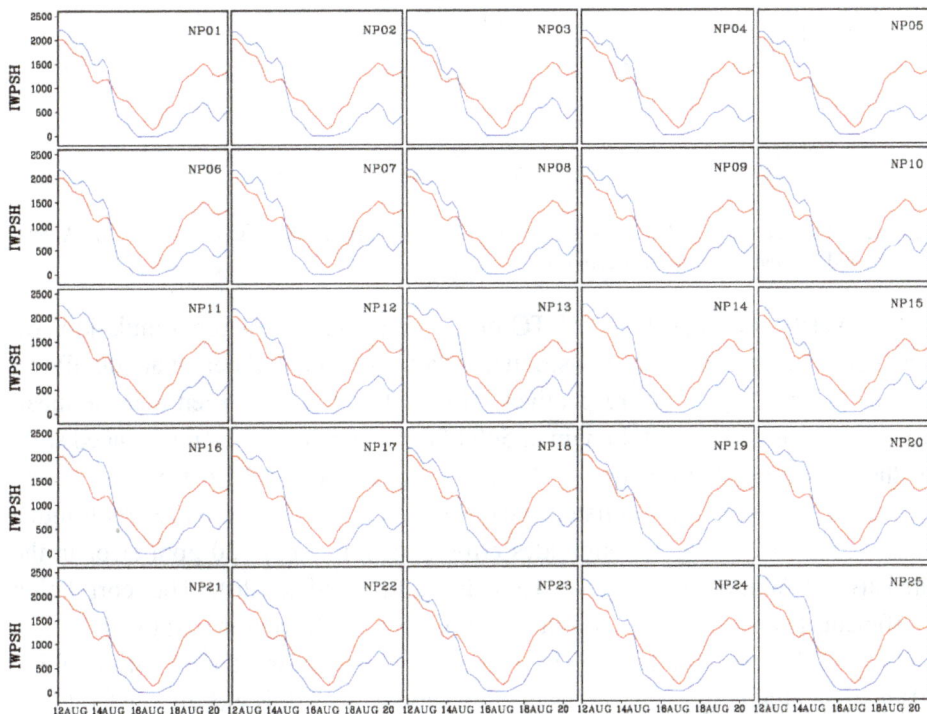

Fig. 2.11. Temporal variations of the observed (solid line) and simulated (dotted line) intensity indices of the western Pacific subtropical high from 00:00 GMT on August 12 to 12:00 GMT on August 20, 1997, for 25 experiments with different configurations of nudging parameters as listed in Table 2.6.

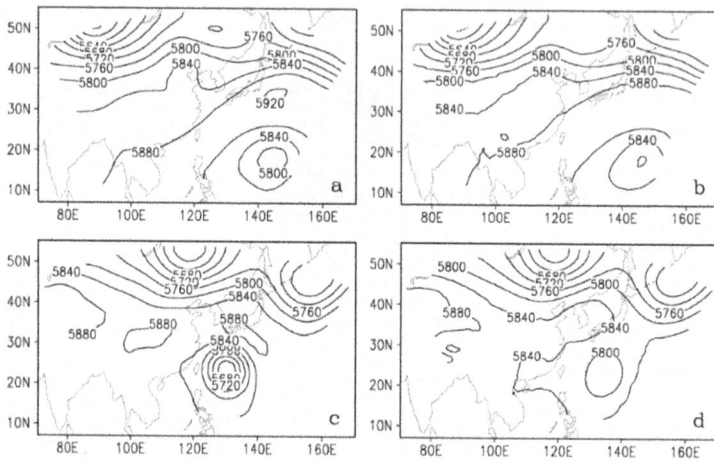

Fig. 2.12. Observed (a, c) and NP19-simulated (b, d) daily geopotential height at 500 hPa on August 12 (a, b) and August 16 (c, d), respectively.

To verify the importance of TC on the regional circulation simulation, we depicted the observed and NP19-simulated daily geopotential height at 500 hPa on August 12 and August 16, respectively (Fig. 2.12). It can be seen that the mean circulation over the model domain at 500 hPa on August 12 was reproduced well, for the well-simulated position of Winnie (Fig. 2.12a and 2.12b). However, the discrepancy between the simulated and observed circulation south to 35°N is much greater for about 5 degree longitudes error in position and 120 gpm error in the intensity of Winnie on August 16 (Fig. 2.12c and 2.12d). The correlation coefficients between observations and simulations for the region east to 100° E and south to 35° N is 0.9694 and 0.7315, respectively. Therefore, the evolution of 500 hPa geopotential height, as well as the intensity of WPSH, is closely related to the TC track, and the simulated TC track error is the primary cause for the failure in simulating east Asian climate in summer [Zhong, 2006].

2.3.3 *Discussions*

TCs over the WNP and landfall in East Asian countries make great contribution to the formation of the regional climate over East Asia, and the regional climate is significantly affected by the frequency of TC activity in terms of weakening the WPSH when TCs are over the WNP and interrupting the summer monsoon when TCs make landfall over China.

The configuration of nudging parameters for the model buffer zone can significantly affects the TC track simulation. Although the best configuration of nudging parameters does not completely eliminate the error in simulating the track, it can largely reduce the errors. Different parameter configuration would generate RMSE in the track simulation by more than two times. Therefore, the appropriate nudging parameter configuration, which maintains the equilibrium between Newtonian term and Laplace diffusion term of the prognostic equations in buffer zone, is crucial for improving the TC track simulation. Neither strong nudging experiment nor weak nudging experiment is beneficial for simulation of the TC track.

To some extent, the error in the simulation of the TC track is related to that of the intensity of WPSH. The weaker simulated WPSH would be the cause of the turning of TCs ahead of time, which will result in great error of track simulations against track observations. Therefore, improvement only with the RCM lateral boundary scheme does not fundamentally eliminate errors in the simulation of the TC tracks. The key challenges in eliminating the errors are to determine how to solve the problem of simulated weaker WPSH, as well as the appropriate presentation of the interaction between TCs and WPSH.

Moreover, the same sets of simulation driven by ERA40 also shows that the appropriate nudging parameter configuration is still to maintain the equilibrium between Newtonian term and Laplace diffusion term of the prognostic equations in buffer zone, and neither strong nudging experiment nor weak nudging experiment is beneficial for simulation of the track.

In addition, the experiments on the impact of model resolution on the TC track simulation suggests that a more reasonable track can be obtained from the simulation at a higher resolution, but the TC is not landed yet as reality for all the experiments, and the simulated intensity is usually much weaker than observed one. It should be pointed out that the improvement of track for the simulation at a higher resolution may come from the positioning of TC partially, therefore, the discrepancy of track can not be eliminated totally for the simulation at higher resolution, and meanwhile, one can not expect the much higher resolution for the climate models, though it has been demonstrated that the model would perform well at high resolution [Giorgi *et al.*, 1996] and its performance is also related to the domain choice [Seth and Giorgi, 1998] and buffer zone size [Zhong *et al.*, 2010].

It has been known that the model physics plays a fundamental role in TC simulation, though it is as yet unclear whether and to what degree the simulated TC track, structure, intensification, and intensity can be affected by using different physics parameterization schemes [Wang, 2002]. For example, the development of TC is sensitive to the transportation of sensible heat, latent heat and momentum in the underlying surface [Braun and Tao, 2000], thus the planetary boundary layer (PBL) scheme will be important in TC simulation. With the MRI mesoscale nonhydrostatic model, it was found that the precipitation structure induced by Typhoon Flo is dependent to the microphysics scheme of the model at a great extent [Murata *et al.*, 2003], which will in turn reflect the track and intensity of TC through feedback and interaction mechanism. Moreover, the simulated tracks of TCs are affected by the detailed microphysics transport in the cumulus parameterization scheme [Hogan and Pauley, 2007]. Therefore, the results for the BZS and nudging parameters should be verified by experiments with different model physical schemes and for other TC cases.

Chapter 3

Impact of Cumulus Parameterization Scheme on the Tropical Cyclone Track

3.1 Introduction

It is well known that the GCMs often generate unacceptable features in simulations of the complex East Asian monsoon, especially in a summer rainfall simulation over eastern China [Yu *et al.*, 2000]. For example, none of the ten GCMs that participate in the Climate Variability and Predictability/Monsoon GCM Intercomparison Project can realistically reproduce the observed Meiyu rainband [Kang *et al.*, 2002]. All the models exhibit limited skills in capturing the summer monsoon rainfall and its variability [Zhou *et al.*, 2008, 2009]. One of the main atmospheric systems affecting the summer precipitation over China is the WPSH. The seasonal extension and withdrawal of the WPSH are responsible for the rainfall distribution over East China; therefore, the simulated precipitation is directly dependent on the capability of the models to capture the activities of the WPSH and the corresponding regional circulations.

As a dynamical downscaling tool of the GCM, RCMs have been developed and applied for climate studies for more than 20 years [Dickinson *et al.*, 1989; Giorgi, 1990; Giorgi *et al.*, 1993a, b; McGregor, 1997; Leung *et al.*, 2003; Wang *et al.*, 2003; Pal *et al.*, 2007; Liang *et al.*, 2012]. Although the ability of RCMs in regional climate simulation is superior to that of the GCMs, there still exists great uncertainties especially over East Asia. This is largely due to the complex topography and land surface conditions, as well as the strong intraseasonal, interannual, and interdecadal variability of the monsoon system [Tao and Chen, 1987; Zou and Zhou, 2013]. Recent studies suggest that the erratic departure of simulated TC tracks from their realistic positions is possibly a primary reason for the RCM's failure in simulating the East Asian summer monsoon [Zhong, 2006; Kubota and Wang, 2009; Cha *et al.*, 2011].

TC track in RCMs is sensitive to CPS [Zhong, 2006]. TC activity is also related to variation in the WPSH, which is either overestimated or underestimated in most RCM simulations [Giorgi *et al.*, 1999; Lee and Suh, 2000]. Large scale forcing plays a crucial role in successful simulations of both the East Asian monsoon climate and TC activities over the WNP [Zhong, 2006]. In addition, RCM simulations are often case-dependent. Despite the great uncertainties in RCM simulations, one distinct feature is that the WPSH can almost always be well simulated when TCs are not active. In contrast, a large bias often appears in WPSH simulations when TCs are active over the WNP [Zhong, 2006; Zhong and Hu, 2007; Fudeyasu *et al.*, 2010].

Although some researchers have found a large bias in the WPSH simulation when TCs are active over the WNP, the involved physical mechanisms have not been revealed. Our overall goal in this study is to explore the physical mechanism behind the current model's failure in simulating the WPSH when TCs are active over the WNP. The objective of this paper is twofold. We will first investigate the sensitivity of TC motion and the WPSH to the CPS and then try to explore the possible mechanism reasons and the involved physical processes. A case study of TC Megi (2010) is performed in this work. Two CPSs are used in the simulation of Megi to investigate TC track sensitivity.

3.2 Sensitivity of TC tracks to the choice of CPS

3.2.1 *Experimental design*

Typhoon Megi is the most intense TC with the longest lifespan over the western North Pacific and SCS in 2010. Megi originated over an area of disturbed weather about 600 km to the east of the Philippine Archipelago around 0000 UTC 12 October 2010 and developed quickly through the day. The Joint Typhoon Warning Center (JTWC) classified the system as a tropical depression at 0900 UTC 13 October. Possibly due to the influence of the WPSH, the depression system moved slowly in a west-northwest direction toward the Philippines. Meanwhile, the depression intensified to a tropical storm around 1200 UTC 13 October. Later on October 14, an eye-like structure of the storm could be seen clearly from satellite images. Japan Meteorological Agency (JMA) upgraded Megi to a severe tropical storm and JTWC upgraded it to a category-1 typhoon. JMA upgraded Megi to a typhoon on October 15 [Kieu *et al.*, 2012; Wang *et al.*, 2013].

The storm initially moved northwestward and then turned west-southwestward along the southern periphery of the WPSH. During this process, it underwent significant intensification due to the highly favorable conditions for tropical storm development. By the time it made its first landfall over the Philippines on 18 October, Megi has become one of the strongest tropical cyclones recorded to make landfall. After crossing the Luzon Island, Megi moved slowly as a trough over central China deepened and extended over the SCS, leading to a break in the subtropical ridge. Due to the strong influence of the trough and the weakening subtropical high over the SCS, Megi experienced a sharp northward turning around 0000 UTC 19 October, then turned north-northeastward. It weakened into a tropical storm and finally a tropical depression when it made its second landfall at Zhangpu in Fujian Province, China on October 23.

The Weather Research and Forecasting model version 3.3 (WRFV3.3), is employed to simulate TC Megi. Initial and boundary conditions are derived from the NCEP final analysis data (FNL, http://dss.ucar.edu/dsszone/ds083.2). A 20-km resolution domain with 36 vertical levels is set up for the simulation of Megi. The model domain is centered at (22°N, 122°E) with 160 (north-south) × 180 (east-west) grid points, including complex topography and land-sea contrast. It extends far enough south to capture Megi's Track. The simulation starts at 0000 UTC, 14 October and ends at 0000 UTC, 24 October, with a total of 240-hour integration that covers the entire life span of Megi.

The physical parameterizations used in this study include the single-moment 3-class microphysics scheme [Hong *et al.*, 2004], and the Mellor-Yamada-Janjić boundary layer scheme [Mellor and Yamada, 1982; Janjić, 2002]. More details of WRF physics and dynamics can be found in Skamarock *et al.* [2008]. Two experiments with different cumulus parameterization scheme, i.e. Grell-Devenyi (GD) cumulus parameterization scheme [Grell and Devenyi, 2002] and Betts-Miller-Janjić (BMJ) cumulus parameterization scheme [Betts and Miller, 1986; Janjić, 1994, 2000], are selected for the simulation of Megi. All other physical schemes and model settings are the same in the two experiments. Results using these two different schemes are compared to investigate sensitivity of TC simulations to different cumulus parameterization schemes.

3.2.2 *Simulation results*

According to the best-track data obtained from JTWC, Megi is characterized by strong intensity, long duration, and fast development, with a typical turning track. It was generated over the WNP (11.9°N, 141.4°E) at 0000 UTC, 13 October 2010. Shortly after its formation, Megi started moving westward and continued to gain strength. Around 0330 UTC, 18 October, Megi made landfall over Isabela Province, Philippine. It became weak when passing Luzon island but rapidly regained strength over the SCS. Late on 19 October, Megi was forced to move northward, then north-northeastward towards a break in the subtropical ridge caused by an approaching mid-latitude trough. It landed over the coastal region of southern China around 0500 UTC 23 October (Fig. 3.1).

Fig. 3.1 shows that Megi's track is well-captured by the GD run. There is no notable difference in the simulated track between the BMJ and GD runs before 0000 UTC 18 October. However, the simulated storm in the BMJ run turns northward earlier than observation, whereas in the GD run it continues to move westward and turns northward over the SCS at about 1200 UTC 19 October. The result of the GD run is consistent to observations.

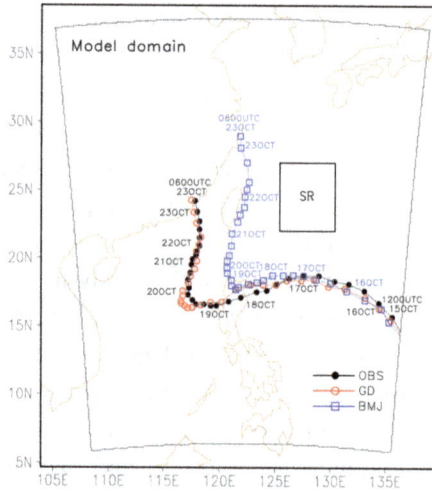

Fig. 3.1. The model domain and simulated storm tracks in the sensitivity experiments with different cumulus parameterization schemes. The observed best track at 6-h intervals (black dotted line) is overlaid.

TCs over the WNP are usually steered by the large-scale environmental flow, especially the flow in the southern edge of the WPSH at 500 hPa [Chan and Gray, 1982; Wu *et al.*, 2005; Zhong, 2006]. In the following part we will discuss the reason why the simulations using different cumulus parameterization schemes generate different TC tracks.

The simulated 500 hPa geopotential height at 0000 UTC 15 October and 18 October using the GD and BMJ schemes are shown in Fig. 3.2. It can be seen that before the simulated storm enters the model domain (15 October), the WPSH is slightly weaker to the east of Taiwan in the BMJ simulation, compared to that in the GD run, since the scope of the geopotential height contour of 5880 m is smaller in the BMJ run than in the GD run (Figs. 3.2a and 3.2b). As the storm continues moving westward, the weakening of the WPSH in the BMJ run becomes more notable prior to the significant departure of the simulated TC track (Figs. 3.2c and 3.2d). Apparently, Megi's unrealistic early turning in the BMJ run is attributed to the weak WPSH, which subsequently causes changes in steering flow.

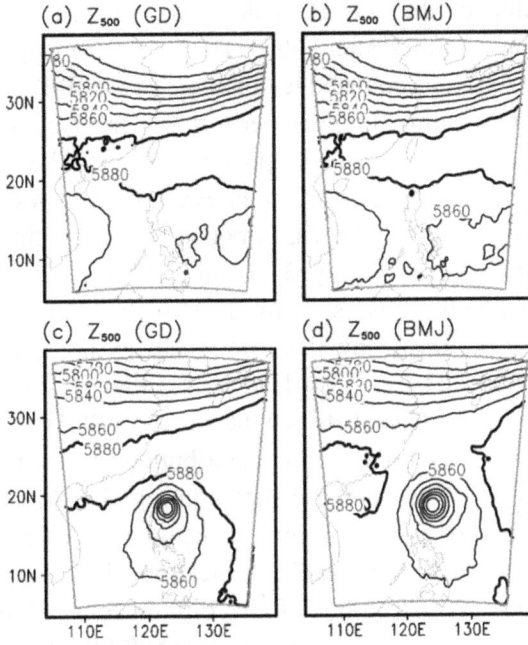

Fig. 3.2. The simulated geopotential height (m) at 500 hPa in the GD (left panel) and BMJ runs (right panel) at 0000 UTC 15 October (a, b) and 18 October (c, d) respectively.

Since the failure in simulating the track of Megi by the BMJ run is directly related to the intensity of the simulated WPSH and its associated steering flow, we define a sensitivity region (SR in Fig. 3.1) (22°N-27°N, 125°E-130°E), where the variation of the WPSH in the BMJ run is significantly different from that in the GD run. The difference of WPSH intensity in the SR between the BMJ and GD runs serves as an important index to assess the failed simulation of TC activity in the BMJ run. Fig. 3.3 shows the time-height cross-sections of differences in geopotential height and temperature, averaged over the SR between the BMJ and GD runs. Terms contributing to temperature tendency [i.e., horizontal advection (HA), vertical transportation (VT), and diabatic heating (DH)] are also shown in Fig. 3.3. The equation for temperature tendency is written as follows.

$$\frac{\partial T}{\partial t} = -\mathbf{V} \cdot \nabla T - \omega(\frac{\partial T}{\partial p} - \frac{1}{C_p \rho}) + \frac{\dot{Q}}{C_P} \tag{3.1}$$

where T is the temperature; ρ is the air density; p is the pressure; ω is the vertical velocity in the pressure coordinates; $\mathbf{V}$ is the horizontal air motion; and $\dot{Q}$ denotes the diabatic heating rate. The three terms on the right-hand side of Eq. (3.1) represent contribution of HA, VT and DH, respectively. As shown in Fig. 3.3, the unrealistic weakening of the WPSH in the BMJ run can be attributed to the abnormal temperature variation after 0000 UTC 17 October (Fig. 3.3b). Compared with that in the GD run, the simulated temperature in the BMJ run is colder in the lower troposphere, but warmer in the upper troposphere. Note that only the BMJ run produces the temperature profile of upper warming and lower cooling. Such a different vertical profile of temperature in the BMJ run cannot change the surface pressure notably, but it causes a decrease in geopotential height at 500 hPa under hydrostatic constraint conditions (Fig. 3.3a and 3.3b) and results in a split of the WPSH, which is eventually responsible for the unrealistic early turning of the TC in the BMJ run (Fig. 3.2d). Meanwhile, differences in the simulated TC activity between the BMJ and GD runs also lead to significant differences in WPSH through interaction and feedback between the TC and WPSH.

The differences in several terms contributing to temperature tendency between the BMJ and GD runs are also shown in Fig. 3.3. Although the effect of HA on temperature tendency is much smaller than that of VT and DH, it makes a notable contribution to differences in the vertical profile of temperature between the BMJ and GD runs. Similar to that of DH, difference in HA between the BMJ and GD runs is positive (negative) above (below) 500 hPa from 1200 UTC 16 to 0000 UTC 18 October, corresponding to the time period when the TC is active in areas south of the SR. The effects of VT and DH on temperature tendency are almost always out of phase, indicating an approximate equilibrium relationship between diabatic heating and adiabatic cooling (warming) due to upward (downward) motion [Mapes and Houze, 1995]. The effect of VT, however, cannot completely offset that of DH on temperature tendency, thus the profile of HA+VT+DH is still consistent with that of DH. The latter is sensitive to the choice of the cumulus parameterization schemes and plays an important role in determining the vertical profile of temperature. DH difference between the BMJ and GD runs is the main reason for warmer air above and colder air below 500 hPa, since it gives a similar pattern to HA+VT+DH in the height-time cross-section (Fig. 3.3e and 3.3f), as well as that of temperature (Fig. 3.3e and 3.3b).

We further examine the diabatic heating rate caused by microphysics ($\dot{Q}_{MP}$), by cumulus scheme ($\dot{Q}_{CP}$), by radiation scheme ($\dot{Q}_{RA}$), and by boundary layer parameterization ($\dot{Q}_{BL}$). Total diabatic heating rate ($\dot{Q}$) can be expressed as:

$$\dot{Q} = \dot{Q}_{MP} + \dot{Q}_{CP} + \dot{Q}_{RA} + \dot{Q}_{BL} \qquad (3.2)$$

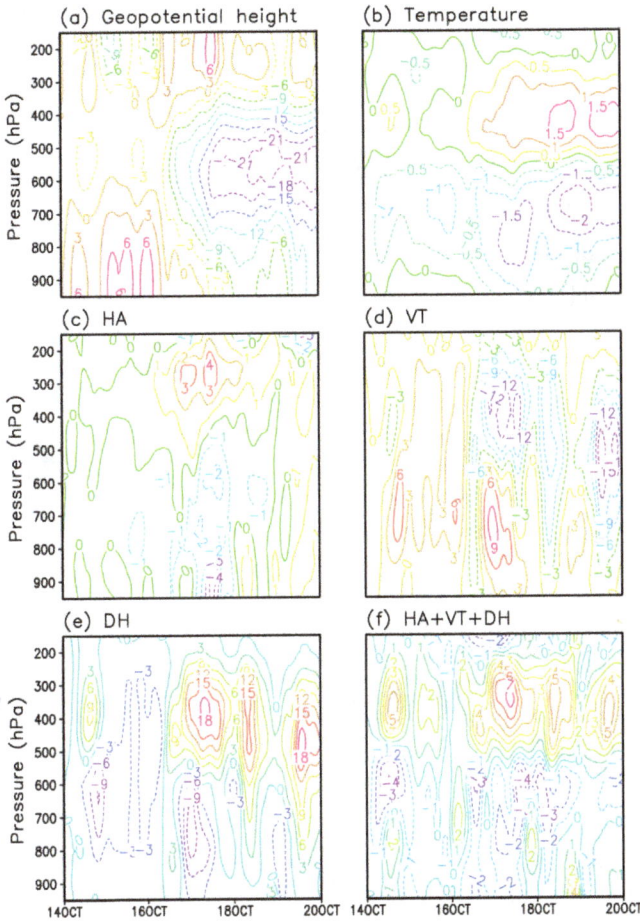

Fig. 3.3. Time-height cross-sections of differences in geopotential height (m), temperature (°C), and terms contributing to temperature tendency (°C d[-1]) between the BMJ and GD runs averaged over SR: (a) geopotential height; (b) temperature; (c) HA; (d) VT; (e) DH; (f) HA+VT+DH.

Differences in each component that contributes to the diabatic heating rate between the BMJ and GD results are shown in Fig. 3.4. The difference in $\dot{Q}_{MP}$ is the largest above 500 hPa and after October 17. Its vertical distribution and temporal variability are quite similar to that of DH difference, despite the slight difference in magnitudes between them (Figs. 3.3e and 3.4a). Above (Below) 500 hPa, $\dot{Q}_{MP}$ can be considered as the primary heat source (sink) that is responsible for the differences in temperature between the BMJ and GD results after 0000 UTC 17 October. Above analyses suggest that, compared to the other three physical schemes (i.e., cumulus, boundary layer, and radiation), microphysical latent heating plays the most important role in determining the differences in DH and temperature between the BMJ and GD results. Hence differences in the microphysics component in the BMJ and GD results directly lead to large discrepancies in Megi and WPSH simulations. As is shown, $\dot{Q}_{CP}$ also contributes greatly to the difference in DH, although the magnitude of the difference is smaller than that caused by $\dot{Q}_{MP}$. It is positive in the levels between 800-500 hPa with a maximum at about 600 hPa and negative in the levels above 500 hPa. This pattern of difference caused by $\dot{Q}_{CP}$ is significantly different from the vertical distributions of DH and temperature differences. However, it does not mean that the cumulus parameterization scheme contributes little to the vertical distributions of DH and temperature differences shown in Figs. 3.3b and 3.3e.

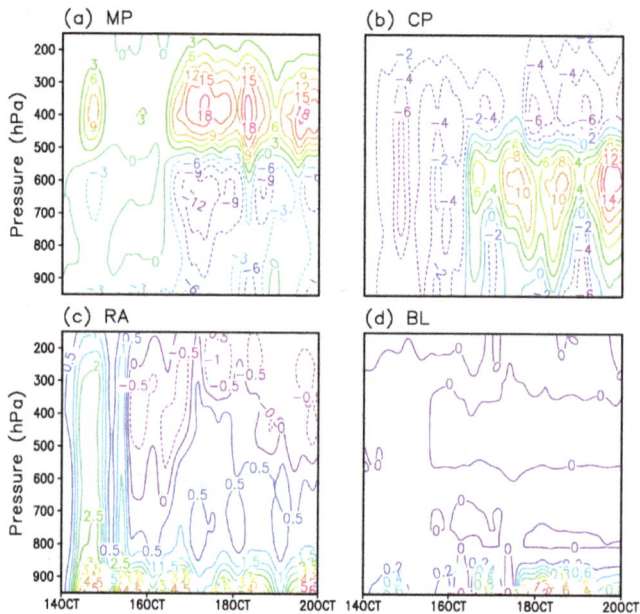

Fig. 3.4. As in Fig. 3.3, but for the differences in each term contributing to DH: (a) microphysics parameterization; (b) cumulus parameterization; (c) radiation parameterization; (d) boundary layer parameterization.

In fact, the CPS contributes to DH not only by directly changing the temperature profile, but also by indirectly changing the moisture profile and thus the microphysical latent heating. Differences between the BMJ and GD results caused by the radiation and boundary layer schemes mostly appear in the boundary layer below 850 hPa and are much smaller than those caused by microphysics and cumulus schemes. Note that, the differences between the BMJ and GD results caused by microphysics, boundary layer and radiation schemes are consequences of using different cumulus schemes, since all physical schemes are the same except cumulus parameterization in our sensitivity experiments.

To understand the source of the difference in microphysical latent heating in SR, we depict the total mixing ratio of model-simulated hydrometeors at 400 hPa for the BMJ and GD cases in Fig. 3.5. The simulated hydrometeors in the GD run are concentrated over a small area near the storm center, while in the BMJ run they cover a much larger area and with higher amount, especially in SR. Namely, the BMJ run produces larger amounts of anvil clouds near the TC region than the GD run. The overestimation of anvil clouds in the BMJ run leads to the bias in vertical temperature simulation (Fig. 3.3b), and eventually affects the simulations of the WPSH and TC motion (Figs. 3.2d and 3.1).

To further investigate the difference over the SR in the time-height cross-section of DH in the BMJ and GD runs, we depict azimuthal-averaged cross sections of hydrometeors for the simulated TC from the center to 800 km in the GD and BMJ runs (Fig. 3.6). The temperature difference between the BMJ and GD runs at 0000 UTC 18 October is also shown in Fig. 3.6. Unlike the GD scheme, which uses ensemble technique, the BMJ scheme imposes an enthalpy adjustment and is very sensitive to deep-layer moisture [Betts and Miller, 1986; Janjić, 1994, 2000]. Due to the sufficient moisture supply over the ocean, the simulated eyewall of Megi and the anvil cloud in the BMJ run is much more abundant than that in the GD run in terms of hydrometeors. This is directly related to the difference of both DH and temperature between the BMJ and GD runs over the SR, suggesting that the rich anvil cloud generated by the BMJ has destroyed the WPSH structure and results in an unrealistic TC track in the BMJ simulation.

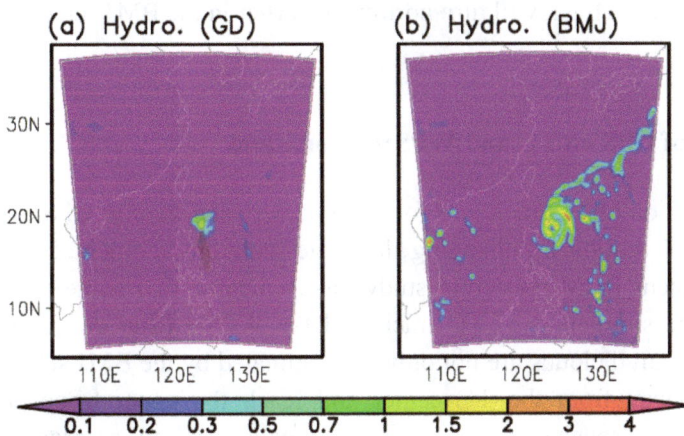

Fig. 3.5. Horizontal distribution of total mixing ratio of model-simulated hydrometeors (g kg^{-1}) at 400 hPa height at 0000 UTC 18 October 2010: (a) the GD case; (b) the BMJ case.

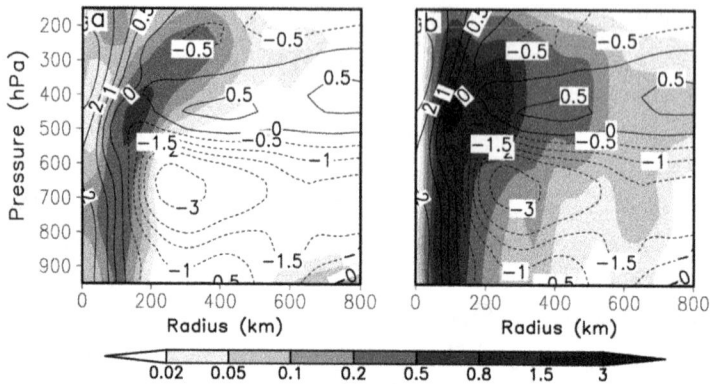

Fig. 3.6. Azimuthal-averaged cross-sections of storm in terms of the mixing ratio of total simulated hydrometeors (g kg^{-1}; shaded) in the GD (a) and BMJ runs (b), and the temperature difference of storm between the BMJ and GD runs (°C; contoured) at 0000 UTC 18 October.

As suggested by Powell [1990], the stratiform precipitation of the storm extends outward as anvil rain, and some of the strong anvil showers in the BMJ run can generate penetrative downdrafts beneath the anvil cloud. This leads to an extended heat source due to condensation in the anvil cloud and an extended heat sink due to precipitation evaporation below the anvil cloud [Willoughby, 1988]. Because of the wider and thicker anvil cloud in the BMJ run, both the warming above and the cooling below 500 hPa are stronger than those in the GD run, resulting in a temperature difference in the SR. Consistent with Stossmeister and Barnes [1992], the changed temperature profile causes a decrease in geopotential height at 500 hPa in the SR (Fig. 3.3a and 3.3b) and a weakening of the WPSH. As a result, the TC track will turn northward earlier in the BMJ run than in the GD run.

3.3 Effects of CPS on TC and WPSH simulations

RCM simulation of TC track was found to be very sensitive to CPS, which could affect the TC motion by changing the large-scale environmental flow [Zhong, 2006]. A recent RCM modeling study also indicated that some CPSs perform poorly in the simulation of TC track and the WPSH [Sun *et al.*, 2014a]. For example, the anvil clouds are not only overestimated by the BMJ scheme but also extend far away from the TC center and reach the upper troposphere. Such unrealistically extensive anvil clouds eventually lead to a large bias in the simulation of the temperature profile. The bias in the temperature profile results in erroneous weakening of the WPSH and thus the early recurvature of the TC. In contrast, both the WPSH intensity and the TC track can be better reproduced in the simulation using the Grell-Devenyi scheme [Grell and Devenyi, 2002].

Although the sensitivity of the WPSH and TC track simulation to CPSs has been investigated under the condition when TCs are active over the WNP, questions do remain. For example, it is unknown whether the erroneous estimation of the cumulus convection in CPSs can be an important reason for the failure in simulating the TC track and WPSH. The best manner in which to identify and correct problems in the CPSs is unknown. Finally, it is critical to link the biases in CPSs to specific physical mechanisms. Here, the BMJ scheme is used to investigate the effects of CPS on simulations of TC and WPSH activities. A suite of sensitivity experiments on the BMJ scheme is designed for the case study of TC Megi (2010). Another case study of TC Songda (2004) is also performed as a comparison with the case of TC Megi.

3.3.1 *Experimental design*

The model configuration parallels that of our previously work [Sun *et al.*, 2015a] as follows in the next two paragraphs. The model used in this study is the Weather Research and Forecasting model version 3.3.1 (WRF V3.3.1) [Skamarock *et al.*, 2008]. The initial and lateral boundary conditions are obtained from the $1° \times 1°$ National Centers for Environmental Prediction (NCEP) final analysis data (FNL) at 6-h intervals (http://rda.ucar.edu/datasets/ds083.2/). A 20-km resolution domain is set up for the simulation of Megi and the WPSH. There are 36 uneven σ levels in the vertical, extending from the surface to the model top at 50 hPa. The model domain is centered at (22°N, 122°E) with 160 (north-south) × 180 (east-west) grid points, which includes complex topography and land-sea contrast. It extends far enough south to allow simulation of the WPSH withdraw and the recurvature of Megi. The simulation is initialized at 0000 UTC 14 October and ends at 0000 UTC 24 October 2010. The 240-hour integration spans most of the life of Megi. The time interval for the output of model results is 1 hour.

The physical parameterizations used are also the same as those in Sun *et al.* [2015b], which include (i) the single-moment 3-class microphysics scheme [Hong *et al.*, 2004]; (ii) the Mellor-Yamada-Janjić boundary layer scheme [Mellor and Yamada, 1982; Janjić, 2002] coupled with the Monin-Obukhov surface layer scheme [Janjić 1996, 2002]; (iii) the Rapid Radiative Transfer Model (RRTM) [Mlawer *et al.*, 1997] for longwave radiation calculation and Goddard scheme [Chou and Suarez, 1994] for shortwave radiation calculation, and (iv) the 5-layer thermal diffusion scheme [Skamarock *et al.*, 2008] for land surface processes.

More details of WRF physics and dynamics can be found in Skamarock *et al.* [2008].

Three experiments with different CPS are conducted in this study. These CPSs are (1) BMJ scheme [Betts and Miller, 1986; Janjić ,1994, 2000], (2) GD scheme [Grell and Devenyi, 2002], (3) KFEX scheme [Kain and Fritsch, 1990; Kain, 2004]. An extra experiment with no CPS (NOCP) is also conducted for the purpose of comparison. All other physical schemes and model settings are the same in the four experiments described above. Results using these different CPSs are compared to investigate the sensitivity of TC and the WPSH simulations to different CPSs.

The main disadvantage of the moist convective adjustment schemes is that the tropical atmosphere does not approach a moist adiabatic equilibrium state in the presence of deep convection. Thus, Betts and Miller [1986] proposed a new convective adjustment scheme that relaxes observed quasi-equilibrium thermodynamic structure. Although the Betts-Miller scheme does not explain the detailed physical interaction between the cloud and its environment, it will give more realistic convective heating and moistening in the vertical because of the use of reference profiles based on observations. Quoting the work of Betts and Miller [1986], the original version of the BMJ scheme was tuned to the Global Atmospheric Research Programme Atlantic Tropical Experiment (GATE) data. The BMJ scheme differs from the Betts-Miller scheme in several important aspects. One of the basic postulates of the BMJ scheme is that the basic features of these regimes can be characterized by a parameter that is called "cloud efficiency" [Janjić, 1994]. This parameter is defined by

$$E = \text{const}_1 \frac{\overline{T} \Delta S}{c_p \sum \Delta T \Delta p} \tag{3.3}$$

Here, const_1 is a nondimensional constant, $\overline{T}$ is the mean temperature of the cloud, Δp are the depths of the model layers in terms of pressure, c_p is the specific heat at constant pressure, and the symbols of ΔS and ΔT denote changes in entropy and temperature within a convection time step Δt, respectively. The cloud efficiency is proportional to a nondimensional combination of the entropy change over the time step, the precipitation over the time step, and the mean temperature of the cloud [Janjić, 2000]. Δq denotes the change in specific humidity within a convection time step Δt. ΔT and Δq are expressed as

$$\Delta T = (T_{\text{ref}} - T'') \frac{\Delta t F(E)}{\tau} \tag{3.4}$$

$$\Delta q = (q_{\text{ref}} - q^n) \frac{\Delta t F(E)}{\tau} \tag{3.5}$$

In equations (3.4) and (3.5) the subscript ref indicates the equilibrium reference profiles [Betts, 1986], the superscript n denotes the values of temperature and specific humidity at the model levels at the beginning of the time step, and τ is a constant relaxation time [Betts, 1986]. Thus, an assumption is made that the convective forcing is proportional to an increasing function of the cloud efficiency $F(E)$.

To further investigate the effects of CPS on simulations of TC activity and the WPSH, we have taken the BMJ scheme as a sample and conducted five sensitivity experiments with different values of the weighting factor (α) in the BMJ scheme. $F(E)$ in the BMJ scheme is weighted by the parameter α, which varies from 2.0 to 0.1 in the five experiments (hereafter B2.0, B1.0, B0.5, B0.2 and B0.1). As can be seen from equations (2) and (3), the parameterized heating and drying rates in the BMJ scheme are also weighted by the parameter α in these sensitivity experiments. Note that the default value of α is 1 in the BMJ scheme, and the aforementioned NOCP experiment can be considered as the sensitivity experiment with $\alpha = 0$. All other physical schemes and model settings are the same in the five experiments.

To verify the performance of the BMJ scheme in simulating the WPSH when TCs are absent over the WNP, we have also conducted an additional suite of experiments with different values of the weighting factor (α) in the BMJ scheme (e.g., 1.0, 0.5, 0.2 and 0.1). These experiments are initialized at 0000 UTC 1 November and run for one month, ending at 0000 UTC 30 November 2010. TCs are absent over the WNP during this period. Except for the different simulation period, the model domain and configuration in this suite of experiments are all the same as that in the former suite.

3.3.2 Simulation results

Fig. 3.7 compares the storm track simulated in the experiments with the JTWC best track of Megi. The model using the GD scheme can well reproduce the track of Megi before and after its turning with an average track error of about 50 km at 6-h intervals, but the simulated track is not very accurate when using the other two CPSs or never using a CPS at all. Specifically, the model with any of the four CPSs performs well in simulating the northwestward movement of TC before 0000 UTC 17 October and the west-southwestward movement of TC along the edge of the WPSH before the TC made landfall in the Luzon Island. However, the model with different CPS produces very different results after 1800 UTC 18 October. In the BMJ and KFEX runs the simulated TC turns northward earlier than observation, while the simulated TC in the GD and NOCP runs continues to move southwestward and turns northward over the seas about 300 km west of the Luzon Island at about 1800 UTC 19 October. These results clearly indicate that the simulated storm track is sensitive to the choice of CPS.

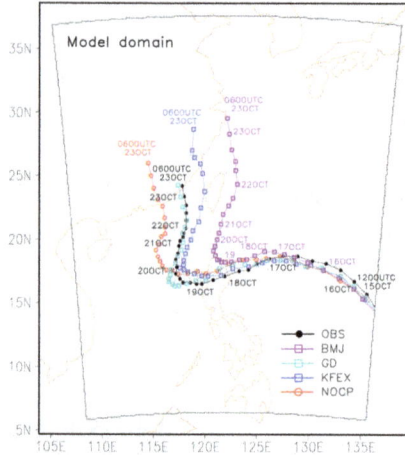

Fig. 3.7. The model domain and simulated storm tracks in the sensitivity experiments with the Betts-Miller-Janjić (BMJ) scheme, the Grell and Devenyi (GD) scheme, the Kain-Fritsch (KFEX) scheme, and none CPS (NOCP) in the case of Megi (2010). The observed best track (OBS) at 6-h intervals (black dotted line) is overlaid.

Fig. 3.8 shows that the simulated storm track is sensitive to the value of α in the BMJ scheme. Decreases in α value from 1.0 to 0.1 in the BMJ scheme delay the northward turning of the simulated storm, resulting in more accurate simulations of the storm track. In contrast, the simulated storm track moves northeastward and deviates from the observation as the α value increases from 1.0 to 2.0. This result implies that the CPS induced heating/drying is overestimated in the BMJ scheme, resulting in a large bias in the simulation of the storm track in B1.0 and B2.0. However, this does not mean that the less heating/drying the CPS generates, the better the forecast is. Comparing the simulated storm track in B0.1 with that in NOCP, one can find that the simulated storm track moves further westward and deviates from the observation as the α value further decreases from 0.1 to 0.0 (Figs. 3.7 and 3.8). In addition, due to the orographic blocking effect of Taiwan's mountain ranges, the simulated storm in B0.5 exhibits a characteristic eastward deflection when the storm approaches Taiwan. It turns to the northwest after passing over the central mountain range of Taiwan, and then resumes a northward track toward China. In other words, the simulated storm in B0.5 turns cyclonically when it passes over Taiwan, which is similar to results found in several previous studies of terrain influence on TC track changes [e.g., Chang, 1982; Bender *et al.*, 1987; Wu, 2001; Lin *et al.*, 2002].

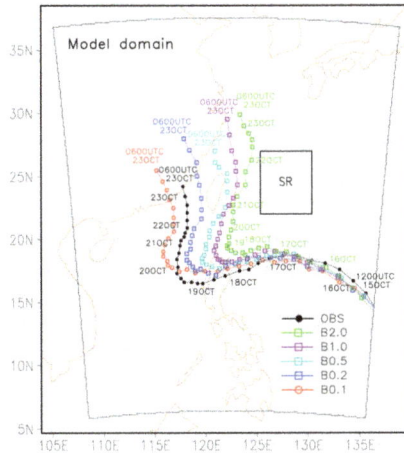

Fig. 3.8. The model domain and simulated storm tracks in the sensitivity experiments with varying α in BMJ scheme in the case of Megi (2010). The observed best track at 6-h intervals (black dotted line) is overlaid.

Because TCs are steered primarily by the large-scale environmental flow, the characteristics of TC tracks over the WNP are modulated by the extension and withdrawal of the WPSH. Fig. 3.9 shows the 500 hPa geopotential height in the sensitivity experiments at 0000 UTC 15 October and at 0000 UTC 18 October. These experiments are initialized at 0000 UTC 14 October. Before the simulated TC enters the model domain (e.g., 0000 UTC 15 October), all experiments give a similar result (Figs. 3.9a-d). As the storm continues moving westward, the simulated WPSH becomes weakened at various degrees in these experiments prior to the significant departure of simulated TC position (Figs. 3.9e-h). A comparison of Figs. 3.8 and 3.9 suggests that, due to the strong influence of the steering flow in the southern edge of the WPSH, the time and location of the northward turning of the storm is closely related to the degree of weakening of the WPSH. Actually, Megi's unrealistic early turning simulated in some sensitivity experiments (e.g., B1.0, B0.5, and B0.2) can be attributed to the unrealistic split of the WPSH. Note that the unrealistic break of the WPSH simulated in these experiments is not caused by the strong storm intensity, since the TC intensity simulated in these experiments except for B0.2 is not stronger than that in B0.1 in terms of minimum surface level pressure.

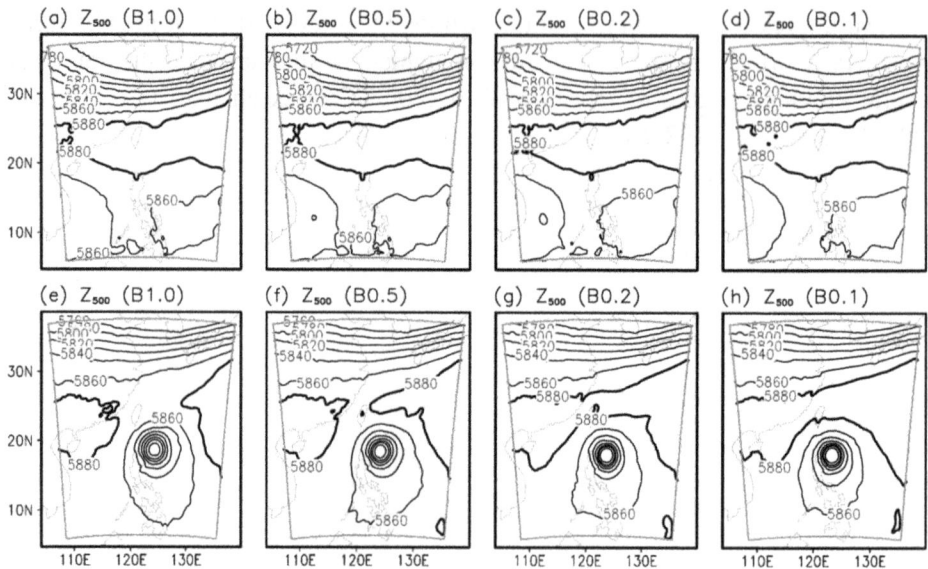

Fig. 3.9. The simulated geopotential heights (m) at 500 hPa in the sensitivity experiments for weighting factor (α) at 0000 UTC 15 October (top panels) and at 0000 UTC 18 October (bottom panels). All simulations are initialized at 0000 UTC 14 October. (a, e: $\alpha = 1.0$; b, f: $\alpha = 0.5$; c, g: $\alpha = 0.2$; d, h: $\alpha = 0.1$).

Similar to the results of Sun *et al.* [2014a, 2015a], the simulated WPSH is also insensitive to the effects of CPS when TCs are absent over the WNP. Fig. 3.10 illustrates the simulated 500 hPa geopotential height at 0000 UTC 30 November in the four experiments that are initialized at 0000 UTC 01 November. The results are verified against the NCEP reanalysis data and show that the simulated synoptic circulation systems are not sensitive to the CPS since all the sensitivity experiments perform well in simulating the WPSH in terms of the 500 hPa geopotential height over the 29-day integration. Results of this series of experiments clearly indicate that the simulated WPSH is not sensitive to the effect of CPS when TCs are absent over the WNP. This further confirms that errors in the WPSH simulation are closely related to errors in the TC simulation. It is important to note that, the unrealistic withdrawal and extension of the simulated WPSH contribute greatly to the failure in RCM simulation of TC motions in this case, while the erratic departure of the simulated TC track from its observed position is possibly a primary reason for the RCM's failure in simulating the WPSH [Zhong, 2006]. Interaction and feedback between the WPSH and TC motion are interwoven in RCM simulations, making it a challenging issue to address the root cause of the large bias in simulations of the WPSH and TC track.

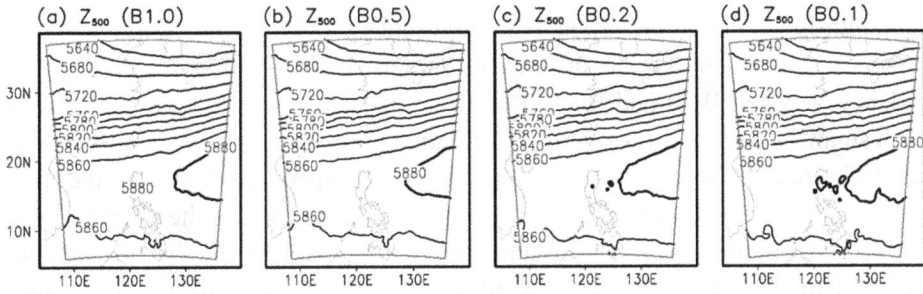

Fig. 3.10. The simulated 500 hPa geopotential height (m) at 0000 UTC 30 November in the four additional experiments. All simulations are initialized at 0000 UTC 01 November.

3.4 Physical mechanism

3.4.1 *Temperature profile*

As defined in Sun *et al.* [2015b], the sensitive region (SR, shown in Fig. 3.8) is over (22°N-27°N, 125°E-130°E), where the variation of the WPSH is significantly different among the sensitivity experiments. The difference in the WPSH intensity in the SR among the sensitivity experiments serves as a useful index to assess the large bias in simulations of the TC track in experiments B2.0, B1.0, and B0.5.

This study provides an in-depth analysis on the difference in temperature among different experiments, finding the source of the difference in temperature among our sensitivity experiments. According to Eq. (5) in Sun *et al.* [2015b], the contribution of individual physical process on the temperature difference among the different experiments can be determined. Fig. 3.11 illustrates not only the pressure-time cross sections of differences in geopotential height and temperature, but also the cross sections of difference in the three terms contributing to temperature tendency [i.e., horizontal advection (HA), vertical transport (VT), and diabatic heating (DH)] averaged over the SR between B1.0 and B0.1 simulations. The break of the WPSH in B1.0 can be attributed to the abnormal temperature structure after 1200 UTC 16 October (Fig. 3.11b). Compared with that in B0.1, the simulated temperature in B1.0 is colder in the lower troposphere below 600 hPa by up to -0.8 K, but warmer in the upper troposphere above 600 hPa by up to 1.6 K. Such a different vertical profile of temperature in B1.0 causes a decrease in geopotential height with the maximum value at about 600 hPa through hydrostatic adjustment (Fig. 3.11a and 3.11b) and results in the break of the WPSH (Fig. 3.9e), which eventually contributes to the unrealistic early turning of the TC in B1.0 (Fig. 3.8). In addition, this pattern of difference in temperature profile between B1.0 and B0.1 only appears when TCs are active over the WNP, and does not appear in the month-long simulations when TCs are absent.

Fig. 3.11 also shows the differences in terms contributing to temperature tendency between B1.0 and B0.1. The effect of HA on temperature tendency is much smaller than that of VT and DH, and thus plays a minor role in determining the difference in the vertical distribution of temperature between the two experiments. The effect of DH on temperature tendency is opposite to that of VT, indicating an approximate equilibrium relationship between the diabatic heating and adiabatic warming (cooling) due to the downward (upward) motion [Mapes and Houze, 1995]. The impact on temperature tendency of VT, however, cannot completely offset that of DH. As a result, the vertical distribution of HA+VT+DH is still basically similar to that of DH. Thereby, it is the difference in distribution of moisture and vertical motion that drives DH and then impacts the vertical temperature profile, and thus is the main reason for the warmer air above and colder air below 550 hPa, which will be further discussed in the following paragraphs. This is clearly reflected in the DH pattern (Fig. 3.11e), which is similar to that of HA+VT+DH (Fig. 3.11f) and temperature (Fig. 3.11b).

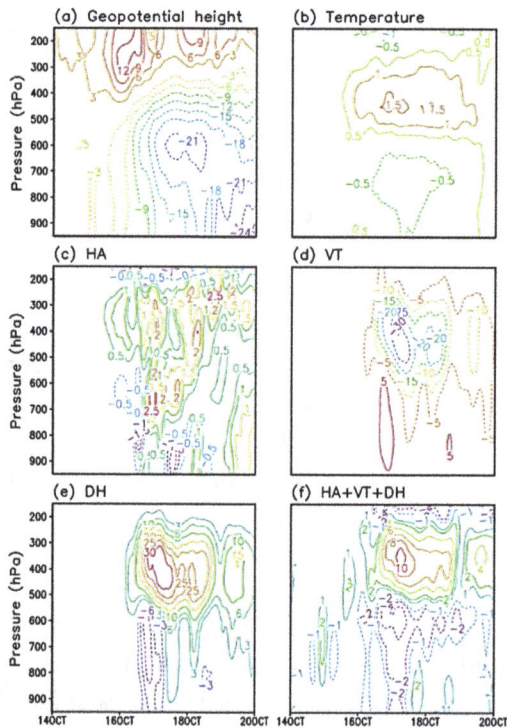

Fig. 3.11. Pressure-time cross sections of differences in geopotential height (m), temperature (K), and terms contributing to temperature tendency (K d^{-1}) between the B1.0 and B0.1 integrations averaged over SR: (a) geopotential height; (b) temperature; (c) HA; (d) VT; (e) DH; (f) HA+VT+DH.

DH can be divided into four components that are caused by microphysics ($\dot{Q}_{MP}$), by the cumulus scheme ($\dot{Q}_{CP}$), by the radiation scheme ($\dot{Q}_{RA}$), and by the boundary layer parameterization ($\dot{Q}_{BL}$), respectively. Differences in each component that contributes to the difference in DH between B1.0 and B0.1 are shown in Fig. 3.12. Above (below) 500 hPa, $\dot{Q}_{MP}$ is the primary heat source (sink) that is responsible for the difference in temperature between the B1.0 and B0.1 after 1200 UTC 16 October. The strong microphysical cooling below 500 hPa and heating above 500 hPa cause the decrease in geopotential height at 500 hPa (Fig. 3.11a) and thus the unrealistic weakening of WPSH in B1.0 (Fig. 3.9e). This result indicates that differences in microphysics component between the two experiments directly lead to large discrepancies in Megi and WPSH simulations. Because the magnitude of $\dot{Q}_{CP}$ is smaller than that of $\dot{Q}_{MP}$ and its profile is substantially different from the vertical distributions of DH and temperature differences, $\dot{Q}_{CP}$ is not the main cause of the large difference in DH and thus the temperature differences between the two experiments shown in Figs. 3.11b and 3e.

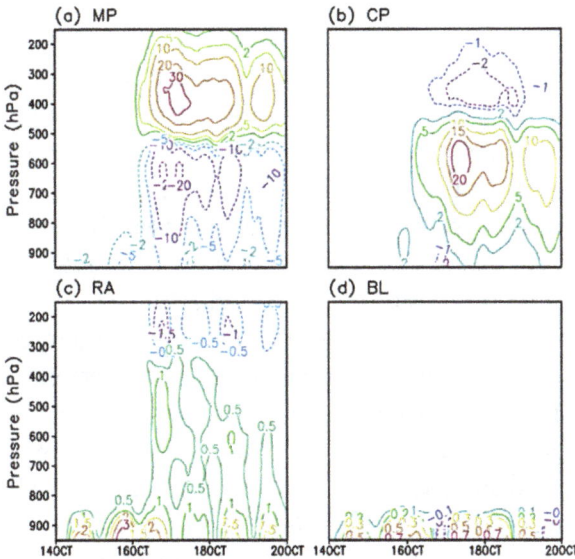

Fig. 3.12. As in Fig. 3.11, but for the differences in each term contributing to DH: (a) microphysics parameterization; (b) cumulus parameterization; (c) radiation parameterization; (d) boundary layer parameterization.

However, this result does not mean that the CPS plays a secondary role in determining the differences in profiles of DH and temperature. In fact, the CPS contributes to DH not only by directly changing the temperature profile, but also by indirectly changing the moisture profile and thus the microphysical latent heating represented by $\dot{Q}_{MP}$. The differences in the other three components (i.e., $\dot{Q}_{MP}$, $\dot{Q}_{RA}$ and $\dot{Q}_{BL}$) are all consequences of the changes in the CPS as the value of α decreases from 1.0 to 0.1 in the BMJ scheme, since all physical schemes are the same except the cumulus parameterization in our sensitivity experiments. Differences in $\dot{Q}_{RA}$ and $\dot{Q}_{BL}$ between B1.0 and B0.1 are much smaller than those in $\dot{Q}_{MP}$ and $\dot{Q}_{CP}$.

3.4.2 *Distribution of the diabatic heating and the hydrometeors*

To examine the source of the difference in microphysical latent heating, we depict the horizontal distribution of DH at different levels (e.g., 300 hPa, 500 hPa, and 700 hPa) for B1.0 and B0.1 in Fig. 3.13. Compared with B0.1, B1.0 produces a larger amount of DH caused by condensation at 300 hPa, which extends far away from the storm center and reaches the SR, leading to the large DH difference between B0.1 and B1.0 in the SR (see also Fig. 3.12a). As the height decreases from 300 hPa to 500 hPa, the DH in B1.0 becomes negative in some regions due to the effect of precipitation evaporation, which leads to a decrease in the DH difference between B1.0 and B0.1 in SR (Figs. 3.13 and 3.12a). For the levels down to 700 hPa, the cooling effect of precipitation evaporation further increases while the warming effect of condensation decreases. As a result, the DH is negative in SR in B1.0, and thus the difference in DH between B1.0 and B0.1 becomes negative at 700 hPa. This is consistent with the large negative value of the DH difference between B1.0 and B0.1 shown in Fig. 3.12a.

Since hydrometeors are important for both the moisture condensation and evaporation, we present in Fig. 3.14 the simulated hydrometeors in B1.0 and B0.1. Compared to that in B0.1, the simulated hydrometeors in B1.0 cover a much larger area far away from the storm center and reach the upper-troposphere, although smaller amounts of hydrometeors are found near the TC eyewall area. Note that the difference in hydrometeors between B1.0 and B0.1 also varies with height in a pattern similar to that of DH difference. Compared with B0.1, B1.0 produces larger amounts of anvil clouds above 500 hPa in terms of hydrometeors in SR. Distribution of the simulated hydrometers is consistent with the spatial pattern of latent heat release caused by condensation as shown in Fig. 3.13. At 700 hPa level, more hydrometeors are generated in B1.0 than in B0.1 in the SR, which is consistent with the larger negative value of DH in B1.0 (shown in Fig. 3.13) that is caused by the effect of precipitation evaporation. Therefore, the large difference in DH between B1.0 and B0.1 in SR can be attributed to the large difference in simulated hydrometeors, which may extend from the TC center and affect large areas far away from the storm.

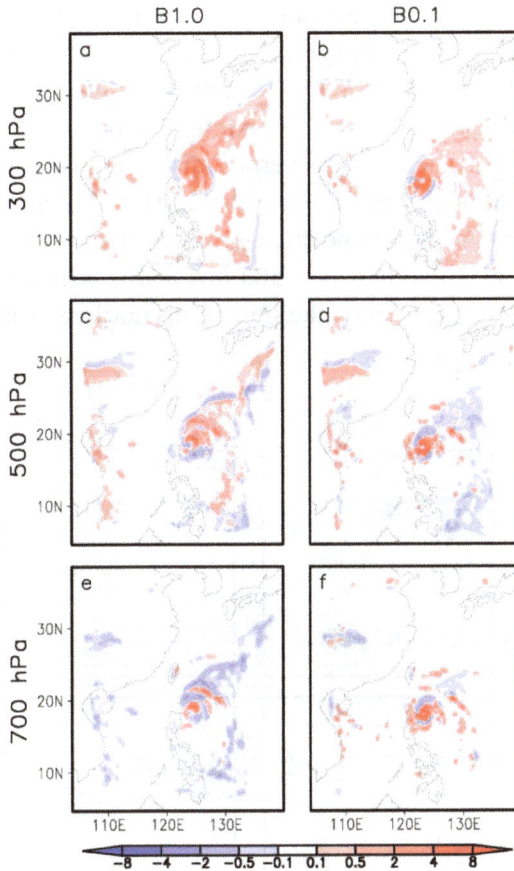

Fig. 3.13. Horizontal distribution of diabatic heating rate (K h^{-1}) at different heights at 0000 UTC 18 October 2010.

The hydrometeors in the model can be divided into cloud and precipitating components, since their contributions to the microphysical latent heating are different. Fig. 3.15 shows the azimuthal-averaged components of cloud and precipitating hydrometers and their contributions to microphysical heating for B1.0 and B0.1. The differences between B1.0 and B0.1 are also shown. The anvil clouds in B1.0 extend to a large area and reach the region far away from the TC center, while the anvil clouds in B0.1 mainly concentrates in a small area near the eyewall (Fig. 3.15a). As suggested by Powell [1990], the stratiform precipitation of the storm extends outward as anvil rain, and some of the strong anvil showers in B1.0 can generate penetrative downdrafts beneath the anvil clouds (Fig. 3.15b).

This leads to a heat source above 500 hPa due to condensation in the anvil clouds and a heat sink below 500 hPa due to precipitation evaporation below the anvil clouds [Willoughby, 1988]. Thereby, it is the cloud (precipitating) component that is responsible for the extra microphysical heating (cooling) in the outer region of the storm above (below) 500 hPa in B1.0 compared to those in B0.1. As the storm approaches SR, the outer region of the storm extends to SR at 0000 UTC 18 October. Results of B1.0 show that the vertical distribution of microphysical latent heating in the SR is basically consistent with that in the outer region of the storm and leads to changes in the temperature profile, which subsequently result in decreases in the geopotential height at 500 hPa and contribute to the unrealistic break of WPSH in B1.0. As a result, the storm turns northward earlier in B1.0 than in B0.1.

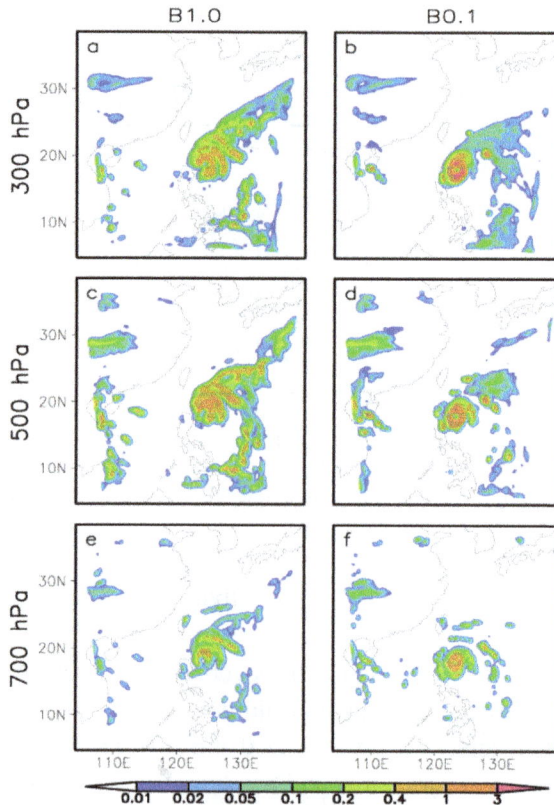

Fig. 3.14. Horizontal distribution of total mixing ratio of model-simulated hydrometeors (g kg^{-1}) at different levels at 0000 UTC 18 October 2010.

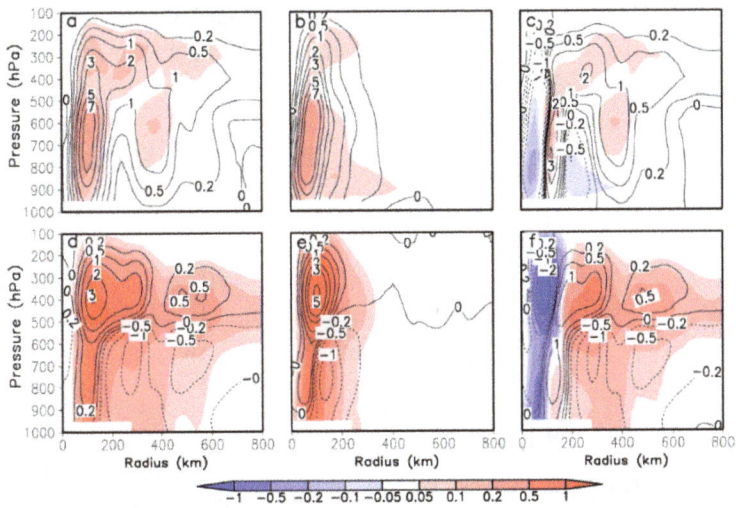

Fig. 3.15. Pressure-radius plots of mean azimuthal cloud (top) and precipitating (bottom) components of microphysical heating (K h^{-1}; shaded) and hydrometeors (g kg^{-1}; contoured) for the B1.0 (a, d), the B0.1 (b, e), and the difference between B1.0 and B0.1 (c, f) at 0000 UTC 18 October.

3.4.3 *Parameterized heating and drying rate*

It is important to note that, the different microphysical behavior between B1.0 and B0.1 is a consequence of different effects of the BMJ scheme, since all physical schemes are the same except for the value of α in the BMJ scheme in the two experiments. In the following paragraph, we will give an explanation for how the CPS difference causes the difference in the simulated microphysical latent heating among our sensitivity experiments. Fig. 3.16 is the same as Fig. 3.15, but for the parameterized heating and drying rates from the BMJ scheme. Following the decrease in the value of α in the CPS, the parameterized heating and drying rates all decrease significantly. Compared with that in B0.1, the parameterized heating in B1.0 is much larger in the layers between 850 hPa and 400 hPa and is basically consistent with the CPS heating in SR (see also Fig. 3.12b). More importantly, the parameterized drying in the outer region of the TC is mainly found in the layers below 500 hPa, while the moistening largely occurs above 500 hPa. As suggested by Grell *et al.* [2013], the CPS drying (moistening) hinders (facilitates) the activation of the microphysics below (above) 500 hPa. This is one primary reason for the generation of extensive grid-resolved anvil clouds in the upper troposphere. This process is responsible for the condensational warming above the freezing level (about 500 hPa) and evaporative cooling below the freezing level, resulting in a negative correlation between the CPS drying and the microphysical latent heating (Figs. 3.12a and 3.16e-10h). Thereby, the CPS affects microphysical latent heating by changing the atmospheric moisture content, and thus contributes to changes in the vertical temperature profile.

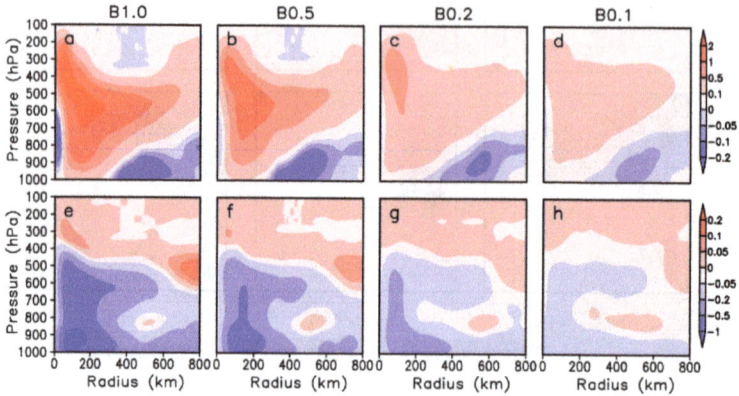

Fig. 3.16. Azimuthal- and time-averaged cross sections of the heating and drying rates from CP at 0000 UTC 18 October: (a-d) heating rate (K h^{-1}); (e-h) drying rate (g kg^{-1} h^{-1}).

3.4.4 *Verification with Songda case*

To further illustrate the impact of the BMJ scheme on TC track and WPSH simulations, we have conducted another case study of TC Songda (2004). Songda (2004) was among the 10 named typhoons that made landfall on the main islands of Japan in 2004 and brought extensive damages to Japan due to strong winds and heavy rainfall [Nakazawa and Rajendran, 2007; Wang *et al.*, 2009]. The sensitivity experiments for TC Songda are the same as those in the case study of Megi (2010), except for different model domain and simulation period. Fig. 3.17 shows the simulated storm tracks, and Fig. 3.18 shows the simulated geopotential height at 500 hPa in these experiments with various values of α at 1200 UTC 3 September in the case of Songda. Similar to results in the case study of TC Megi, the simulated storm track is also sensitive to the value of α in the BMJ scheme in the case study of Songda. Decreases in the value of α from 2.0 to 0.1 in the BMJ scheme delays the northward turning of the simulated storm. The simulated storm continues to move westward, following a track that is more consistent with observation especially after 0000 UTC 05 August. Comparing Figs. 3.17 and 3.18, we can find that differences in the simulated storm track between these sensitive experiments are closely related to differences in the simulated WPSH.

    As the α value decreases from 1.0 to 0.1 in the BMJ scheme, the weakening of the WPSH becomes less significant, and a blocking ridge appears to the north of Songda, delaying the northward turning of the storm. Results for the case study of Songda are consistent with the aforementioned results for the case study of Megi. Both case studies provide evidence to support our findings and conclusions discussed in this paper. Note that the sensitivity of the WPSH simulation to the BMJ scheme may be determined by many factors, such as the distance between the WPSH and the specific storm, the intensity of the WPSH, the storm size, the amount of simulated hydrometeors nearby the storm, etc. Among these factors, storm size may play an important role in determining the level of the sensitivity. We will further discuss the issue in the following paragraph.

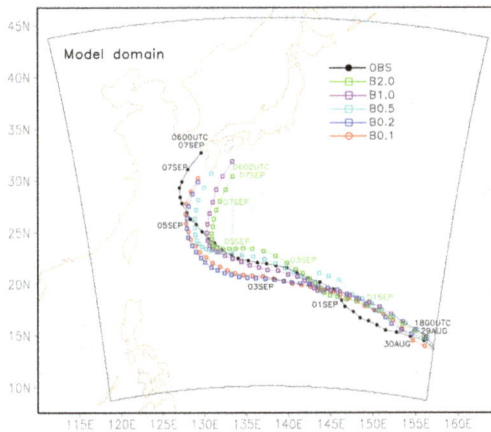

Fig. 3.17. The model domain and simulated storm tracks in the sensitivity experiments with varying α in BMJ scheme in the case of Songda (2004). The observed best track at 6-h intervals (black dotted line) is overlaid.

In an operational setting, storm size is described by the area of the outermost closed isobar (ACI). To compare the simulated storm size of Songda with that of Megi, we have shown the temporal evolution of ACI for Megi and Songda in Fig. 3.19. It clearly demonstrates that the ACI for Megi is highly sensitive to the α value in the BMJ scheme, and decreases significantly as the α value decreases. However, the ACI is notably less sensitive to the α value in Songda than that in Megi. The decreasing trend of ACI for Songda is not very remarkable as the value of α decreases except when α decreases from 1.0 to 0.5. Due to the low sensitivity of the simulated storm size to the α value in Songda case, the anvil clouds, which are influenced by the storm size, are also less sensitive to the α value in Songda case compared to that in Megi case. As described in the previous sections, since the amount of the anvil clouds has smaller changes in Songda than in Megi as the value of α decreases, the corresponding changes in temperature profile, 500 hPa geopotential height and thus storm track are also smaller in the case of Songda than in the case of Megi.

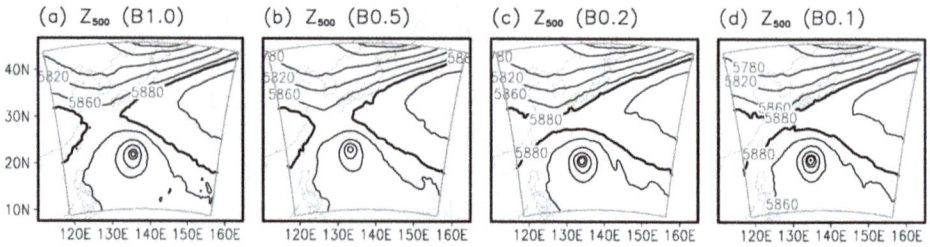

Fig. 3.18. The simulated geopotential height (m) at 500 hPa at 1200 UTC 3 September 2004 in the case study of Songda.

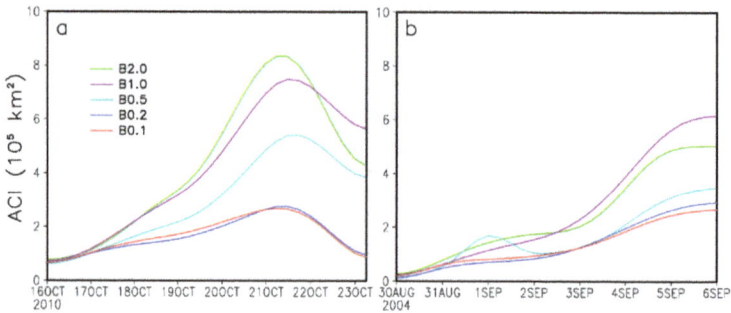

Fig. 3.19. Temporal evolutions of ACI in the sensitivity experiments (a) in Megi case and (b) Songda case.

It is worth noting that the simulated storm track and the WPSH are less sensitive to the weighting parameter in the BMJ scheme in the case study of Songda than they are in the case study of Megi. Further analysis reveals that the storm size plays an important role in determining the sensitivity of the WPSH simulation to the BMJ scheme when TCs are active over the WNP. The low level of sensitivity in Songda case may be eventually attributed to its small size as well as the less sensitivity of storm size to the weighting parameter in the BMJ scheme.

3.5 Discussions

The experiment with the GD scheme generates a realistic track of Megi, which is consistent to that shown by the Best Track data. The experiment with the BMJ scheme, however, gives an unrealistic early recurvature of Megi after its landfall in Philippines. The comparison between the BMJ and GD simulations illustrates how different cumulus schemes affect simulations of TC activity over the WNP for this case. In general, the experiment using the BMJ scheme generates unrealistically abundant hydrometeors in the eyewall of Megi and extensive anvil clouds over areas far away from the TC center. Areas with anvil clouds even extend to the upper troposphere over the WPSH, resulting in a large bias in microphysical latent heating. This bias leads to a warming of the upper troposphere due to condensation in the anvil clouds, and a cooling of the lower troposphere due to precipitation evaporation below the anvil clouds. As a result, the WPSH weakens and the large-scale steering flow becomes anomalously northward, leading to an early recurvature of Megi.

To further address the effects of CPS on WPSH and TC activities, the sensitivity of TC track and WPSH simulation to a specific parameter in the BMJ scheme has been studied using the WRF model. The simulated WPSH strengthens as the CPS induced heating/drying decreases, following the decrease of the value from 2.0 to 0.1 in the BMJ scheme. The decrease in delays the northward turning of the simulated storm due to changes in the large-scale steering flow at the southern edge of the WPSH, and result in a more realistic simulation of the storm track that is consistent with observation. The comparison between B1.0 and B0.1 simulations illustrates how the BMJ scheme affects simulations of the WPSH and TC activity over the WNP for this case. Compared to B0.1, B1.0 produces a stronger drying below 500 hPa and a stronger moistening above 500 hPa, which hinders (facilitates) the activation of the microphysics below (above) 500 hPa. Thus, B1.0 generates extensive anvil clouds in the upper troposphere. These anvil clouds extend far away from the TC center and reach the upper troposphere over the WPSH. Compared with that in B0.1, the overestimated anvil clouds in B1.0 cause stronger condensational warming above the freezing level (about 500 hPa) and evaporative cooling below the freezing level, resulting in a large bias in microphysical latent heating. This bias leads to a warming in the upper troposphere due to condensation in the anvil clouds, and a cooling in the lower troposphere due to precipitation evaporation below the anvil clouds. As a result, the WPSH weakens and the large-scale steering flow becomes anomalously northward, leading to an early recurvature of Megi.

Note that this is a case study. Results of this study demonstrate the importance of a correct representation of anvil clouds in simulating the WPSH and TC track. The erroneous estimation of anvil clouds is a root cause of the large bias in simulating both the WPSH and TC motion at least for the cases in this study. It could also be a reason for errors in many RCM studies, although this requires confirmation using additional models and for more case studies. The findings of this study have shed light on the role of convection parameterization in the simulation of the WPSH and TC, and thus may be conducive to resolving the difficult problem in operational forecast of the WPSH and TC track. This study also implies that, errors in simulating TC motion caused by the unrealistic withdrawal (extension) of the WPSH can be reduced by revising the CP scheme in the numerical model. It should be noted that only the sensitivity of model simulations to a specific weighting parameter in the BMJ scheme has been tested in this study. A weakness of the present study is that the conclusions are based on a single model study for two individual cases, and thus requires confirmation using additional models and performing more case studies. This will be a topic for our future research.

Chapter 4

Feedback of Tropical Cyclone Activities on the Western Pacific Subtropical High

4.1 Introduction

Despite the significant progress in regional climate modeling studies during the past 20 years or so, RCMs still exhibit relatively low skills in simulating the regional climate in the tropics, especially in simulations for the East Asian summer monsoon system [McGregor, 1997; Wang and Wang, 2001; Zhong, 2006; Zhou *et al.*, 2008, 2009]. One of the primary reasons responsible for this failure is that convection in the tropics cannot be well represented in current RCMs. Previous studies have suggested that the RCM simulation of monsoon circulation is very sensitive to different CPS [Zhang, 1994; Leung *et al.*, 1999; Lee ad Suh, 2000; Sun *et al.*, 2014a]. In fact, various microphysics parameterization scheme (MPS) also have a significant impact on convection through microphysical heating/cooling. While great efforts have been taken to investigate the impact of CPS, the sensitivity of the simulated monsoon circulation to MPS schemes, which are equally important, is neglected.

Recent studies indicate that the erratic departure of the simulated TC track from its realistic position is possibly a primary reason of RCM's failure in simulating East Asian summer monsoon [Zhong, 2006; Kubota and Wang, 2009; Cha *et al.*, 2011]. The unrealistic position of the simulated TC over the WNP can lead to errors in simulation of the extent and intensity of the WPSH, and a subsequent failure in East Asian summer monsoon simulation. Despite the consistent and substantial improvements in TC track forecasts, large errors still exist and are of great concern given the overall improvement and the societal expectations accompanying such improvement. Numerous studies have examined various factors, including environment wind errors and storm structure errors that may lead to inaccurate TC position forecast in numerical models. Carr and Elsberry [2000] found that errors in the simulated TC track could be attributed to an unrealistic description of interaction between the tropical cyclone and mid-latitude system. Their distance and spatial scales may significantly affect their interaction. Wang and Holland [1996] suggested that the diabatic heating (DH) could affect the TC motion through downward penetrating flows that are associated with the anticyclonic potential vorticity (PV) anomalies aloft, which are continuously generated by DH. Asymmetric divergent flows associated with convective asymmetries within the vortex core region also contribute to the downward penetrating flows.

While the TC position forecast "busts" can be related to errors in the structure and intensity of the TC vortex [McTaggart-Cowan *et al.*, 2006], errors in the environmental wind appear to be dominant on errors in the TC track forecast [Galarneau and Davis, 2013]. For example, in the forecasts of TC Ike (2008) based on three operational global models, the environmental wind field in all three models steered TC Ike into southern Texas instead of recurving it over the Gulf of Mexico [Brennan and Majumdar, 2011]. This error in the forecasted TC position was attributed to an excessive zonal elongation of the subtropical anticyclone over the southern United States, which induced a more easterly steering flow over the Gulf of Mexico. This error in the structure of the subtropical ridge could be traced back to errors in the environmental initial condition of the model [Komaromi *et al.*, 2011]. As suggested by McGregor [1997], however, the mid-latitude systems are generally reproduced well by RCMs due to the strong large-scale forcing and few cumulus convections. Thus, the relationship between the TC motion and mid-latitude systems (e.g., subtropical high) is an interesting issue that needs to be addressed.

While numerous studies have demonstrated the impact of microphysical scheme on hurricane intensity forecast [Lord *et al.*, 1984; Wang, 2002; McFarquhar *et al.*, 2006; Zhu and Zhang, 2006], little is known about its impact on TC track forecast. Fovell and Su [2007] and Fovell *et al.* [2009] argued that, without imposed large-scale flow, various cloud microphysics and cumulus schemes in a single model could produce large TC movement deviations that could finally lead to significantly different track simulations. The large TC movement deviations are caused by the direct or indirect impact of the microphysical assumptions on storm structure.

In this Chapter, our overall goal is to explore the physical mechanism behind the RCM's failure in simulating the WPSH when TCs are active over the WNP. We will first investigate the sensitivity of TC movement and WPSH to MPSs and then try to reveal the possible reasons and the involved physical processes.

4.2 Model configuration and experimental design

The model used here is the WRF V3.3.1 [Skamarock *et al.*, 2008]. The initial and lateral boundary conditions are obtained from the $1° \times 1°$ NCEP final analysis data (FNL) at 6-h intervals. A 20-km resolution domain is set up for the simulation of

Megi and WPSH. There are 36 uneven σ-levels extending from the surface to the model top at 50 hPa. The model domain is centered at (22°N, 122°E) with 160 (north-south) × 180 (east-west) grid points, including complex topography and land-sea contrast. It extends far enough south to allow capture of the withdrawal of WPSH and the turning of Megi. The simulation starts at 0000 UTC, 14 October and ends at 0000 UTC, 24 October 2010, with a total of 240-hour integration that covers the entire life span of Megi. The time interval for the output of model results is 1 hour.

The physical parameterizations used in this study include (i) the Grell-Devenyi (GD) cumulus parameterization scheme [Grell and Devenyi, 2002]; (ii) the Mellor-Yamada-Janjić boundary layer scheme [Mellor and Yamada, 1982; Janjić, 2002] coupled with the Monin-Obukhov surface layer scheme [Janjić, 1996, 2002]; (iii) the Rapid Radiative Transfer Model (RRTM) [Mlawer *et al.*, 1997] for longwave radiation calculation and Goddard scheme [Chou and Suarez, 1994] for shortwave radiation calculation, and (iv) the 5-layer thermal diffusion scheme [Skamarock *et al.*, 2008] for land surface processes. Four experiments with different MPSs are conducted in this study. All other physical schemes and model settings are the same in the four experiments. Results using these four different MPSs are compared to investigate the sensitivity of TC and WPSH simulations to different MPS scheme. Moreover, to verify the performance of different MPs in simulating the WPSH when TCs are absent over the WNP, we have also conducted an additional suite of experiments using the four MPS schemes. These experiments start at 0000 UTC, 1 November and end at 0000 UTC, 30 November 2010 with a total of 1-month integration. TCs are absent over the WNP during this period. Except for the different simulation period, the model domain and configuration in the four additional experiments are all consistent with that in the former experiments.

The MPSs used in this study include (1) the single-moment 3-class (WSM3) [Hong *et al.*, 2004], (2) Lin [Lin *et al.*, 1983; Chen and Sun, 2002], (3) the single-moment 6-class (WSM6) [Hong and Lim, 2006], and (4) Thompson [Thompson *et al.*, 2004; Thompson *et al.*, 2008]. More specifically, the WSM3 scheme predicts three categories of hydrometers: vapor, cloud water/ice, and rain/snow. This is a so-called simple-ice scheme. A major difference between the WSM3 and other approaches is that the diagnostic equation used for calculation of ice number concentration is based on ice mass content rather than on temperature. It follows Dudhia [1989] to assume cloud water and rain for temperatures above freezing, and cloud ice and snow for temperatures below freezing. The Lin scheme includes six classes of hydrometeors: water vapor, cloud water, rain, cloud ice, snow, and graupel. All parameterization production terms are based on Lin *et al.* [1983] and

Rutledge and Hobbs [1984] with some modifications, including a saturation adjustment following Tao *et al.* [1989] and an ice sedimentation. The scheme is taken from the Purdue cloud model, and the details can be found in Chen and Sun [2002]. The WSM6 scheme is similar to the WSM3 simple ice scheme. However, vapor, rain, snow, graupel, cloud ice, and cloud water are held in six different arrays. Thus, it allows the existence of supercooled water and a gradual melting of snow falling below the melting layer. Some of the graupel-related terms follow Lin *et al.* [1983], but its ice-phase behavior is much different due to the changes of Hong *et al.* [2004]. The Thompson scheme is a new bulk microphysical parameterization (BMP) that has been developed for use with WRF or other mesoscale models. Unlike any other BMP, the assumed snow size distribution depends on both ice water content and temperature and is represented by a sum of exponential and gamma distributions. Furthermore, snow assumes a non-spherical shape with a bulk density that varies inversely with diameter as found in observations. This is different to almost all other BMPs that assume a spherical snow with constant density. More details of the four MPSs can be found in Skamarock *et al.* [2008].

4.3 Simulation results with different MP schemes

Zhong [2006] suggested that the RCMs can realistically reproduce the regional circulation systems (e.g., WPSH). However, they cannot perform equivalently well when TCs are active over the WNP. Cha *et al.* [2011] proposed that typhoons are responsible for the significant systematic errors in long-term regional climate simulations over East Asia due to their impact on large-scale environment. For the purpose of this study, we discuss the performance of RCM under two conditions: when TCs are absent and when TCs are active over the WNP. Fig. 4.1 shows the geopotential height in NCEP reanalysis data and the simulated geopotential height at 500 hPa at 0000 UTC 30 November in the four additional experiments, which are initialized at 0000 UTC 01 November. The results are verified against NCEP reanalysis data. It is found that the simulated WPSH is not sensitive to MPSs since all the sensitivity experiments perform well in simulating the WPSH in terms of geopotential height contour of 5880 m after 29-day of integration. Results of this suite of experiments clearly indicate that the simulated WPSH is not sensitive to MPSs when TCs are absent over the WNP.

Next, we will investigate the sensitivity of WPSH intensity and TC track to the MPSs when TCs are active over the WNP. Previous studies suggested that the large-scale environmental flow played a critical role in determining TC motion. Without imposed large-scale flow, however, a self-propagating vortex motion can be modulated distinctly by microphysics [Fovell *et al.*, 2009]. Fovell and Su [2007] showed that various cloud microphysics assumptions, together with CPS, can have a great impact on hurricane tracks simulated in a regional-scale model at 30-km horizontal resolution, and the impact is quite significant even in relatively short range (2-day) forecasts. Sun *et al.* [2014a] demonstrated that, even with different CPS in the sensitivity experiments, the change of microphysical latent heating is still largely responsible for the discrepancies in the WPSH and TC motion simulations.

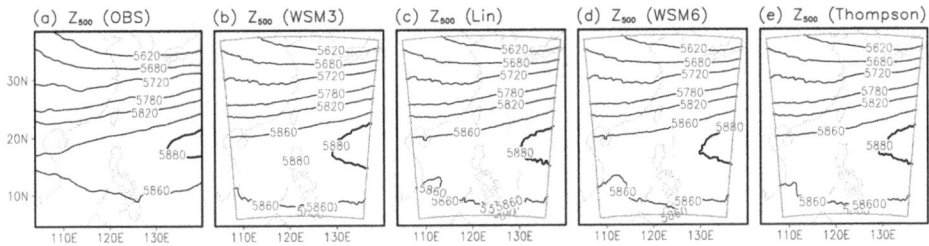

Fig. 4.1. (a) The geopotential height (m) in NCEP reanalysis data and (b, c, d, e) the simulated geopotential height (m) at 500 hPa at 0000 UTC 30 November in the four additional experiments. All simulations are initialized at 0000 UTC 01 November.

Fig. 4.2 compares the storm tracks simulated in the sensitivity experiments with the JTWC best track. The model with WSM3 can well reproduce the track of Megi before and after its turning with an average track error of about 50 km at 6-h intervals, but it performs not so well with the other three MPS schemes. Here, we use control run (CTR) to refer to the WSM3 experiment and sensitivity run (SER) to refer to experiment with Lin scheme, WSM6 scheme and Thompson scheme, respectively. All experiments realistically simulate the northwestward movement of Megi before 0000 UTC 17 October and the west-southwestward movement along the southern periphery of the WPSH until the storm crossed the Luzon Island. Large differences between results of the four experiments occur after 1800 UTC 18 October. The simulated storm in SERs turns northward earlier than observation, whereas in CTR it continues to move westward and turn northward over SCS at about 1800 UTC 19 October. Apparently, the simulated storm track is sensitive to the choice of MPS.

It has been pointed out that the TC track over the WNP is mainly determined by two factors. One is the environmental flow, especially the steering flow in the south of the WPSH at 500 hPa [e.g., Chan and Gray, 1982; Wu *et al.*, 2005; Zhong, 2006], and the other is the dynamic and thermodynamic structure of TC itself [e.g., Holland, 1983; Fiorino and Elsberry, 1989; Wu and Wang, 2000]. Therefore, theoretically the MPS has two ways to affect the TC motion. First, the MPS can modulate the TC motion by influencing the dynamic and thermodynamic structure of TC. Second, the MPS can affect the steering flow of a TC by influencing the extension and withdrawal of the WPSH. In the following section, we will discuss which one is responsible for the failed simulation of TC motion in SERs.

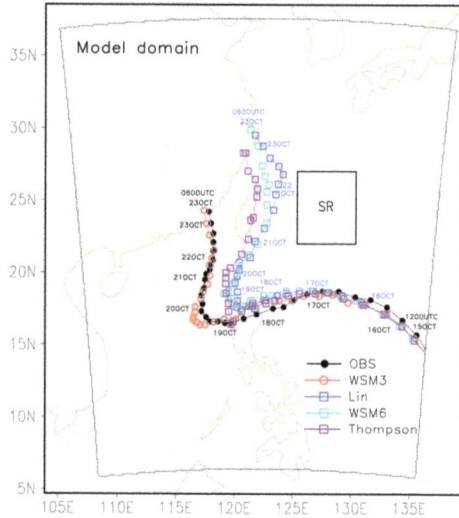

Fig. 4.2. The model domain (sector area) and storm tracks overlaid with the observed best track at 6-h intervals for the sensitivity experiments with different MPS schemes. (SR indicates the sensitivity region for analysis later).

The potential vorticity tendency (PVT) diagnosis technique [Chan, 1984; Wu and Wang, 2000] is utilized to estimate contributions of the TC structure and environmental flow to TC motion. Based on in their simulation of a TC on a β plane without basic flow, Wu and Wang [2000] suggested that the motion of the modeled TC follows the maximum local PVT with azimuthal wavenumber 1 (WN1) in the mid-troposphere. The PVT is given by the sum of the contributions from the horizontal advection (HA), vertical transportation (VT) and diabatic heating (DH). The influence of physical processes on the vortex motion at each level can be identified by the PVT diagnosis approach. As suggested by Wu and Wang [2000], the velocity of the TC motion is estimated based on PVTs in Eq. (4.1) and the individual contributions of various terms in Eq. (4.2) can be estimated in the pressure coordinates.

$$\left(\frac{\partial P}{\partial t} \right)_1 = -\mathbf{V}_{PV} \cdot \nabla P_S \qquad (4.1)$$

$$\left(\frac{\partial P}{\partial t}\right)_1 = \Lambda_1\left[-\mathbf{V}\cdot\nabla P - \omega\frac{\partial P}{\partial p} - g\nabla_3\cdot\left(-\frac{\dot{Q}}{C_p\pi}\mathbf{q} + \nabla\theta\times\mathbf{F}\right)\right] \tag{4.2}$$

where P is PV, P_S is the symmetric component of PV, $\mathbf{V}$ is the horizontal air motion, $\mathbf{V}_{PV}$ is the vortex motion speed estimated from the WN1 component of the PVT, $\mathbf{q}$ is the three-dimensional absolute vorticity vector, p is the pressure, Λ_1 denotes an operator to obtain the wavenumber one component, ∇_3 is the three-dimensional gradient, θ is potential temperature, ω is the vertical velocity in the pressure coordinates. $\dot{Q}$ and $\mathbf{F}$ denote diabatic heating rate and friction, respectively. We apply Eq. (4.1) to each grid point (denoted by subscript i). This will result in a set of linear algebraic equations,

$$\left(\frac{\partial P}{\partial t}\right)_{1i} = -c_x\left(\frac{\partial P_S}{\partial x}\right)_i - c_y\left(\frac{\partial P_S}{\partial y}\right)_i \tag{4.3}$$

The zonal (c_x) and meridional (c_y) components of the velocity of the vortex motion at each level will be determined from equation (4.3). Considering a specified domain in the vicinity of the TC center which is within a radius of 300 km from the TC center, we use the least square method to estimate c_x and c_y by minimizing:

$$\sum_{i\leq N}\left[\left(\frac{\partial P}{\partial t}\right)_{1i} + c_x\left(\frac{\partial P_S}{\partial x}\right)_i + c_y\left(\frac{\partial P_S}{\partial y}\right)_i\right] \tag{4.4}$$

where N is the number of total grid points in the specified domain. The steering by environmental flow is included in HA, which is the first term on the right-hand side of Eq. (4.2), while VT (the second term) and DH (the third term) mainly depend on the dynamic and thermodynamic structure of the simulated TC that is sensitive to the MPS scheme.

Fig. 4.3 shows the temporal variation of the vertically averaged TC speed calculated from the PVT equations as well as its individual contributions of HA, VT and DH in the WSM3 (CTR) and WSM6 experiments (SER). The time period is from Oct. 17 to Oct. 21, which corresponds to the period before and after the time when the TC track simulation between the CTR and SER becomes significantly different. Due to the various vertical extents of the positive PV anomalies in CTR and PR, the mean speed is calculated from 800 hPa to 500 hPa. The results indicate that the mean speed of TC estimated from the WN1 component of the PVT ($\mathbf{V}_{PV}$) is almost identical to that calculated from the time-dependent variational TC center position ($\mathbf{V}_C$) in most cases. Thus, the PVT diagnosis approach is the same efficient in estimating the motion of TC in real TC case as it is in the ideal case in Wu and Wang [2000].

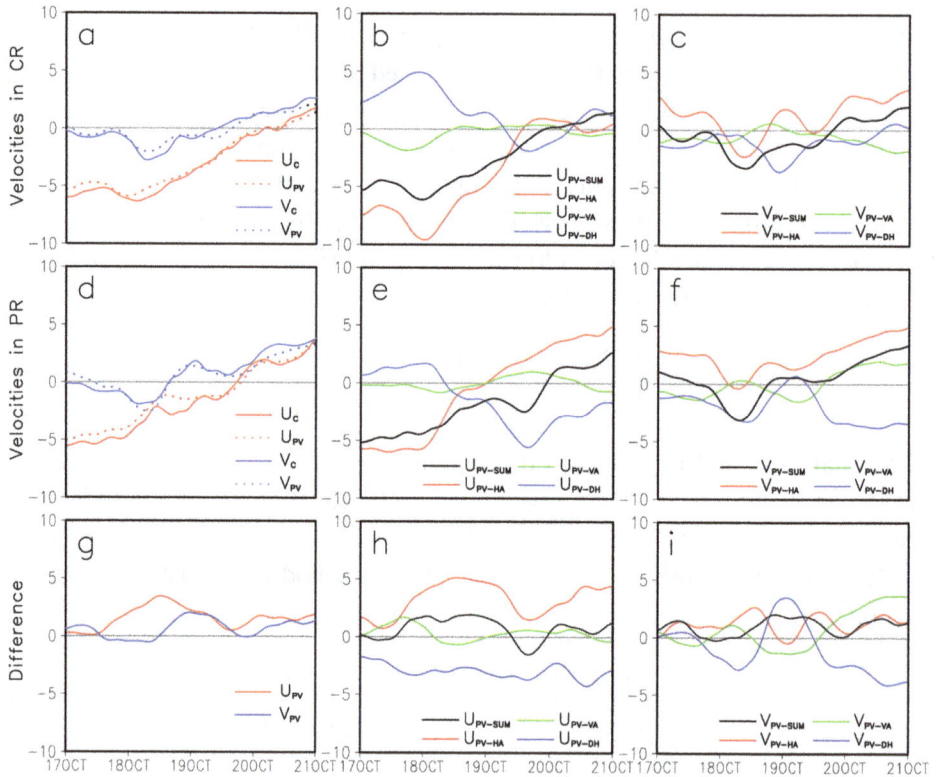

Fig. 4.3. Vertical mean zonal and meridional TC motion speed (m s^{-1}) calculated from the center position (V$_C$), the PV tendency (V$_{PV}$), and individual contributions of HA, VA and DH (V$_{PV-HA}$, V$_{PV-VA}$ and V$_{PV-DH}$) in CTR (top panels), SER (middle), and their difference (SER-CTR, bottom panels). All calculations are averaged within a radius of 360 km from TC center and between the levels 800 hPa and 500 hPa.

According to the Eq. (4.2), the contribution of individual physical process on the difference in TC motion between the CTR and SER can be determined. In the following analysis, we focus on the period before 0000 UTC 19 October, when the difference in TC position is insignificant. This is because, once the difference in TC position becomes significant, the contribution of each single term in Eq. (4.2) may be affected by the difference in the surrounding environment and thus cannot be used to fully explain the difference in TC motion. Due to the zonal extension of WPSH and its strong large-scale forcing in the CTR, contribution of HA is larger than that in SER (i.e., **V**$_{PV-HA}$), leading to the difference in TC zonal motion (Figs.

4.3b and 4.3e). Although in this case the meridional wind is weaker than zonal wind, it still contributes greatly to the TC motion along the meridional direction (Figs. 4.3c and 4.3f). Moreover, the magnitude of zonal $\mathbf{V}_{PV-HA}$ in CTR is much larger than that in SER by up to 4 m s^{-1}, but the magnitude of meridional $\mathbf{V}_{PV-HA}$ in the SER is notably larger than that in the CTR by up to 2 m s^{-1}, especially before 1800 UTC 18 October when the forecast track difference between the CTR and SER is not significant. This indicates that $\mathbf{V}_{PV-HA}$ slows down the westward moving of TC but accelerates its northward moving in the SER. Compared to the contribution of $\mathbf{V}_{PV-HA}$, $\mathbf{V}_{PV-VT}$ contributes little to the TC motion before 0000 UTC 19 October. Hence it is not the direct and major reason for the difference in TC motion (i.e., $\mathbf{V}_{PV}$) between the two experiments. As suggested by Wu and Wang [2000] and Peng *et al.* [1999], due to a fast adjustment between the relatively asymmetric flow (HA) and the asymmetric diabatic heating (DH), the temporal variations of $\mathbf{V}_{PV-HA}$ and $\mathbf{V}_{PV-DH}$ contributions are out of phase. Although $\mathbf{V}_{PV-DH}$ contributes little to the difference in zonal TC motion under the condition of strong zonal steering flow (i.e., zonal $\mathbf{V}_{PV-HA}$), it plays an important role in determining the difference in meridional TC moving since the meridional steering flow is relatively weak. This is consistent with the findings of Wang and Holland [1996]. The contribution of DH, however, is smaller than that of HA in the difference in meridional TC motion, except for a short period approximately from 1800 UTC 18 October to 0600 UTC 19 October when the forecast TC track difference between CTR and SER is significant (Fig. 4.3i). Therefore, it is the contribution of HA ($\mathbf{V}_{PV-HA}$) that is responsible for the difference, especially TC's northward-turning ahead of time in SER. The MPS directly affects TC motion by changing the environmental flow near the TC (e.g. HA in PVT). The environment flow is closely related to the intensity and extent of WPSH. Moreover, by comparing the result using the WSM3 with that using the other two MPSs (i.e., Lin and Thompson scheme), we can reach similar conclusions.

Fig. 4.4 shows the geopotential height in NCEP reanalysis data and the simulated geopotential height at 500 hPa with four MPSs at 0000 UTC 15 October and 18 October. It can be seen that before the simulated storm entering the model domain (e.g., 0000 UTC 15 October), all experiments give the similar WPSH in terms of geopotential height contour of 5880 m (Figs. 4.4b-e), which is also basically consistent with that in NCEP reanalysis data (Fig. 4.4a). Namely, the impact of different MPS is not significant on 15 October, as the starting date is 14 October. As the storm continues moving westward, the simulated WPSH in SERs is greatly weakened prior to the significant departure of the simulated TC track (Figs. 4.4h-j). At 0000 UTC 18 October, the WPSH in SERs collapses and breaks down into two components near Taiwan. This causes substantial changes in the synoptic steering flow and strongly affects the TC motion. As is shown in Figs. 4.1 and 4.4, results of these experiments clearly indicate that the simulated WPSH is sensitive to the choice of MPSs only when TCs are active over WNP. This is consistent with the previous studies [Zhong, 2006; Kubota and Wang, 2009; Cha *et al.*, 2011]. According to the geopotential height at 500 hPa in NCEP reanalysis data (Fig. 4.4f), the WPSH does not break near Taiwan in terms of the geopotential height contour of 5880 m, which is basically consistent with the results in CTR (Fig. 4.4g) but substantially different from the results in SERs (Figs. 4.4h-4.4j). Therefore, the break of WPSH near Taiwan in SERs at 0000 UTC 18 October is unrealistic. Compared with the better model performance in simulating the WPSH when TCs are absent over WNP, we find that the unrealistic break of WPSH in SERs is closely related to the unrealistic simulation of TC motion (or track) when the TC is active over the WNP (Figs. 4.1, 4.2 and 4.4).

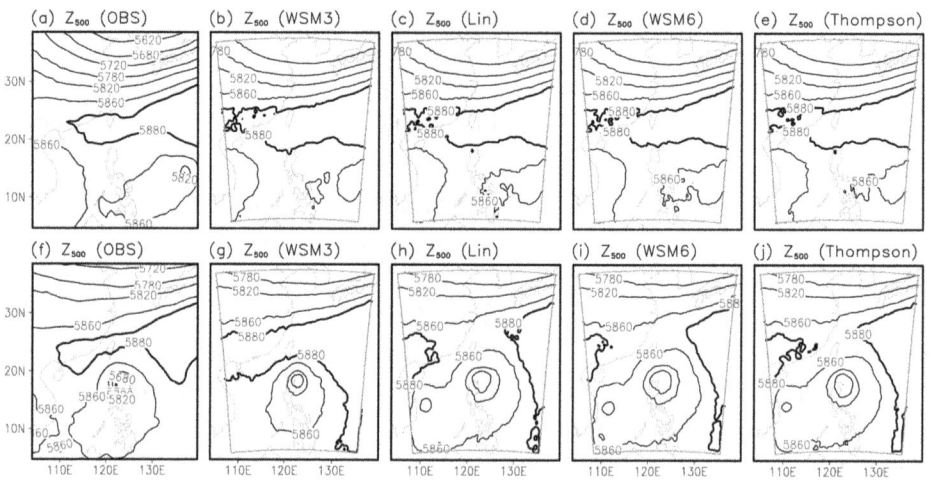

Fig. 4.4. The geopotential height (m) in NCEP reanalysis data and the simulated geopotential height (m) at 500 hPa in the four sensitivity experiments at 0000 UTC 15 October (a, b, c, d, e) and 18 October (f, g, h, i, j), respectively.

Meanwhile, note that there is an additional disturbance at 500 hPa near 15°N, 110°E that is not present in the CTR (Figs. 4f-4h). It could be the result of the initial conditions and microphysics parameterization. However, the impact of the disturbance on the large-scale steering flow and thus the storm track in SERs is very limited for the following three reasons: (1) as suggested by Brand [1970], the rotation of the binary system is sharply dependent upon separation distance when the distance is smaller than 12 latitudes. At 0000 UTC 18 October, the distance between the two storms is larger than 12 latitudes. Such a long distance makes it impossible for the two storms to interact with each other; (2) the difference between the intensity of the two storms is huge. Thus, the impact of the weaker storm on the stronger storm is very limited; (3) most importantly, due to the long distance, the fictitious disturbance is unlikely to exert any impact on the WPSH in SERs, which determines the large-scale steering flow and thus TC track. Apparently, the unrealistic break of WPSH in SERs cannot be attributed to the small fictitious disturbance at 0000 UTC 18 October prior to the significant departure of the simulated TC track.

As mentioned above, we find that the unrealistic positions of TC over the WNP are responsible for the failure in simulating the extent and intensity of WPSH. Note that, this statement is not contradicting with the earlier argument that the TC track over the WNP is mainly determined by the steering flow of WPSH. In this study, as the TC track over the WNP is mainly determined by the steering flow of WPSH, the unrealistic break of WPSH causes errors in the simulation of TC motion, and thus unrealistic positions of the simulated TC in SERs. Subsequently, as the simulated TC position deviates from the observation, errors are also found in the WPSH simulation in SERs. The feedback loop between the inaccurately simulated TC position and the WPSH eventually leads to large biases in simulations of WPSH and TC.

It is suggested that an inappropriate MPS is among those factors that are responsible for the failure in TC motion simulation. The MPS influences the environmental flow of TC through its impact on HA. The unrealistic change in the steering flow along the southern edge of the WPSH is the primary reason for the early turning of TC in SERs. Sensitivity tests can simulate the WPSH quite well when TCs are absent (Fig. 4.1), implying that the failure in WPSH simulations in SERs may be related to the unrealistic description of convections in TC [McGregor, 1997]. However, as shown in Fig. 4.4, the WPSH has broken in the sensitivity region (SR, shown in Fig. 4.2) when the center of SR is within a radius of about 500 km of the TC center at 0000 UTC 18 October. The question how TC convections can lead to the break of WPSH in SR from such a distant place is not answered yet. To address this question, in next section we will discuss the variation of WPSH with respect to the MPS and investigate how the MPS affects the DH in TC and leads to the break of WPSH subsequently.

4.4 Relationship between the WPSH and TC track

4.4.1 *Temperature profile*

Since the failure in simulation of Megi track in SERs is directly linked to the unrealistic break of the WPSH and associated errors in steering flow, we define a sensitivity region (SR, shown in Fig. 4.2) covering (22°N-27°N, 125°E-130°E), where the variation of WPSH in SERs is significantly different from that in the CTR. The difference in WPSH intensity over the SR simulated by the SERs and CTR serves as an important index to evaluate model performance in simulation of the WPSH.

Detailed analysis reveals that the unrealistic break of WPSH in SERs can be attributed to the bias in simulation of temperature profile, which affects the geopotential height and thus the intensity of WPSH. The equation for temperature tendency is written as follows.

$$\frac{\partial T}{\partial t} = -\mathbf{V} \cdot \nabla T - \omega(\frac{\partial T}{\partial p} - \frac{1}{C_P \rho}) + \frac{\dot{Q}}{C_P} \qquad (4.5)$$

where T is the temperature, and ρ is the air density. The three terms on the right-hand side of Eq. (4.5) represent contribution of HA, VT and DH, respectively.

Fig. 4.5 shows the time-height cross-section of the differences in geopotential height and temperature averaged over the SR between results of the SER with WSM6 and CTR with WSM3. Terms contributing to temperature tendency difference (i.e., HA, VT, and DH) are also shown in Fig. 4.5. The unrealistic split of WPSH in PRs can be attributed to the large bias in temperature variation after 0000 UTC 17 October when the storm is approaching SR (Fig. 4.5b). Compared with that in CTR, the simulated temperature in SERs is colder in the lower troposphere below 500 hPa by up to -2.7°C, but warmer in the upper troposphere above 500 hPa by up to 2.2°C. As suggested by Pauley and Smith [1988], such a difference in temperature profile in the SER may not lead to notable changes in the surface pressure, but it causes a decrease of geopotential height at 500 hPa with a magnitude of -29 m under hydrostatic constraint conditions (Figs. 4.5a and 4.5b) and results in the shrink of WPSH areal extent (Fig. 4.4i), which is eventually responsible for the unrealistic earlier turning of TC in SER (Fig. 4.2).

The differences in terms contributing to the temperature tendency between the CTR and SER are also shown in Fig. 4.5. In contrast to its great contribution to TC motion, the contribution of HA to temperature tendency is much smaller than that of VT and DH. Its contribution to differences in the temperature profile between CTR and SER, however, cannot be neglected. Consistent with that of temperature, difference in HA between the two experiments is positive (negative) above (below) 500 hPa from 1200 UTC 16 to 0000 UTC 18 October, corresponding to the time period when TC is active in areas south of SR (Fig. 4.5c). Interestingly, the effects of VT and DH on temperature tendency are always out of phase. As suggested by Mapes and Houze [1995] and Zhang *et al.* [2002], it indicates an approximate equilibrium relationship between the diabatic heating and adiabatic cooling (warming) caused by upward (downward) motion (Figs. 4.5d and 4.5e).

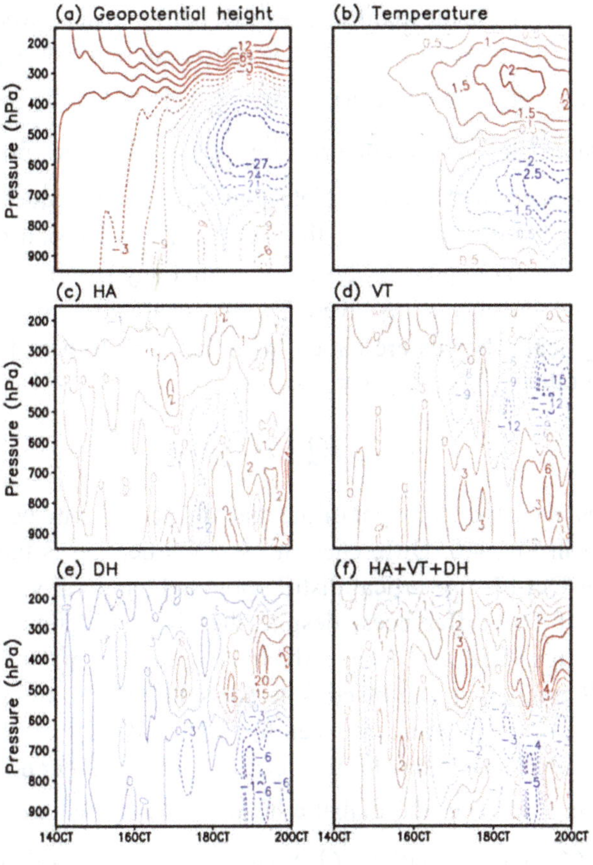

Fig. 4.5. Height-time cross-sections of difference in geopotential height (m), temperature (°C), and the terms contributing to temperature tendency (°C d^{-1}) between SER and CTR averaged over SR. (a) geopotential height; (b) temperature; (c) HA; (d) VT; (e) DH; (f) HA+VT+DH.

However, due to the large magnitude of DH, the effect of VT cannot completely offset that of DH on temperature tendency. Thus, the profile of HA+VT+DH is still roughly similar to that of DH, the latter is sensitive to the choice of MPS and contributes greatly to differences in the temperature profile between CTR and SER (Figs. 4.5e and 4.5f). Apparently, the DH difference between the two experiments is a major reason that causes the different temperature profiles shown in Fig. 4.5b. Temperature is warmer (colder) in the SER than in the CTR above (below) 500 hPa, which is a dividing line that separates upper level temperature response to MPS from that in lower level. The MPS in WRF affects the temperature profile by influencing the profiles of VT and DH. Large biases in VT and DH eventually lead to the unrealistic break of WPSH and thus the earlier turning of TC.

To further investigate the main source of DH difference between the SER and CTR, we examine the diabatic heating rates caused by radiation parameterization ($\dot{Q}_{RA}$), by boundary layer parameterization ($\dot{Q}_{BL}$), by cumulus parameterization ($\dot{Q}_{CP}$), and by microphysics parameterization ($\dot{Q}_{MP}$), which are ranked according their calling sequence in the WRF. Note that the calling sequence for various physical schemes has no impact on the results. This is because in WRF, the physical variables are not updated immediately after any specific physical scheme is called. Only the tendencies are saved. The physical variables are updated only after all the physical schemes are called, then move on to next time step. Total diabatic heating rate ($\dot{Q}$) can be expressed as:

$$\dot{Q} = \dot{Q}_{MP} + \dot{Q}_{CP} + \dot{Q}_{RA} + \dot{Q}_{BL} \qquad (4.6)$$

Differences in each component of the diabatic heating rate between SER and CTR are shown in Fig. 4.6. Difference in the MPS heating is the largest above 500mb and after Oct.17. Its vertical distribution and temporal variability is quite similar to that of DH differences, despite the slight difference in magnitude between them (Figs. 4.5e and 4.6a). Above (Below) the 500 hPa geopotential height, the MPS component can be considered as an important heat source (sink) for the difference in temperature between the SER and CTR after 0000 UTC 17 October. This result clearly indicates that, compared to the other three components of diabatic heating rate (i.e. the diabatic heating rate caused by cumulus ($\dot{Q}_{CP}$), boundary layer ($\dot{Q}_{BL}$) and radiation ($\dot{Q}_{RA}$) parameterizations), microphysics latent heating ($\dot{Q}_{MP}$) plays the most important role in determining the difference in DH and temperature. Large positive (negative) values of difference in both DH and temperature between the CTR and SER appear above (below) 500 hPa, which is a dividing line between different response to MPSs in upper and lower atmosphere (Fig. 4.6a). Hence difference in microphysics scheme in the CTR and SER directly leads to a large discrepancy in Megi and WPSH simulation. Note that, the differences between SER and CTR caused by cumulus, boundary layer and radiation are a consequence of using different MPSs and play a secondary role in determining the difference in DH, since all physical schemes are the same except microphysics in our four experiments. Specifically, although $\dot{Q}_{CP}$ difference is notably weaker than $\dot{Q}_{MP}$ difference between the SER and CTR, it contributes to the warming in the upper and middle troposphere and cooling in the low troposphere especially around 0000 UTC 19 October (Fig. 4.6b). While it is true that $\dot{Q}_{RA}$ difference is much weaker than $\dot{Q}_{MP}$ difference, there is up to 2°C difference in the upper-troposphere (around 200-300 hPa), which is consistent with the magnitude in temperature changes related to cloud-radiation interaction simulated by Jin *et al.* [2014]. This relative small change could have substantial impact on the environment given their wider coverage (Fig. 4.6c). $\dot{Q}_{BL}$ difference is much weaker than $\dot{Q}_{MP}$, $\dot{Q}_{CP}$ and $\dot{Q}_{RA}$ differences, and mainly concentrates in the boundary layer with little effects above (Fig. 4.6d).

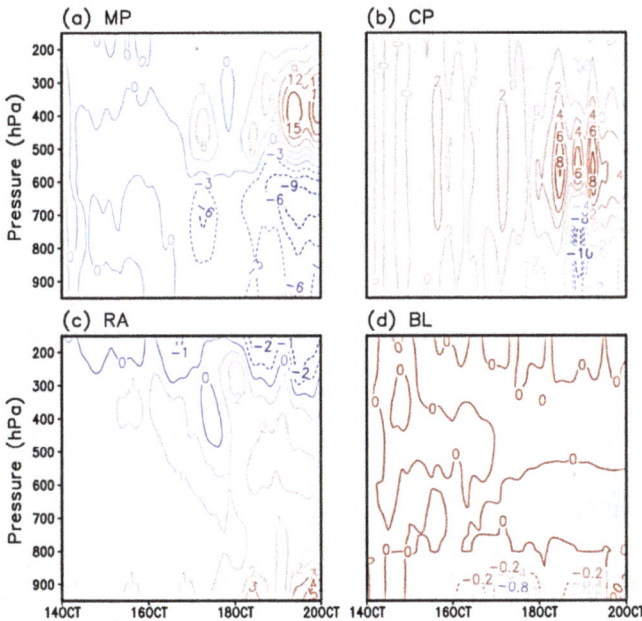

Fig. 4.6. As in Fig. 4.5, but for the terms contributing to DH difference (°C d^{-1}) between SER and CTR. (a) microphysics parameterization; (b) cumulus parameterization; (c) radiation parameterization; (d) boundary layer parameterization.

However, it does not mean that the other three schemes (i.e., cumulus, boundary layer and radiation schemes) contribute little to the vertical distributions of DH and thus temperature differences. According to the results in the CPS sensitivity experiments in Sun *et al.* [2014a], the CPS contributes to DH not only by changing the temperature profile in terms of $\dot{Q}_{CP}$, but also by changing the moisture profile and thus the microphysical latent heating in terms of $\dot{Q}_{MP}$.

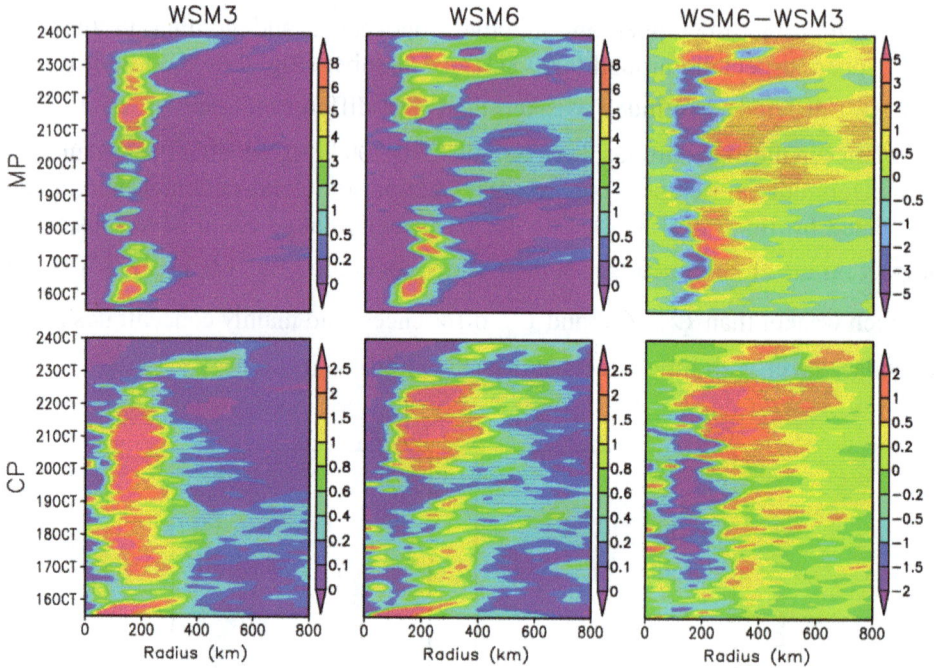

Fig. 4.7. Hovmöller diagram of the azimuthal-averaged diabatic heating from MPS and CP (°C h^{-1}) at 500 hPa for the WSM3 case, the WSM6 case, and the difference between WSM6 case and WSM3 case.

Due to the importance of the diabatic heating caused by MPS and by CPS, we further investigate the role of MPS and CPS on temperature in the storm. Fig. 4.7 provides a Hovmöller diagram of the azimuthal-averaged diabatic heating from MPS and CPS at 500 hPa for the WSM3 case, the WSM6 case, and the difference between WSM6 case and WSM3 case. CPS heating has the similar varying trend as MPS heating. In WSM3 and WSM6 cases, although the magnitude of CPS heating is much smaller than that of MPS heating, the extension of CPS heating is much larger than that of MPS heating. Thus, similar to the effect of MPS heating, CPS heating also has great influence on convection near the storm. More importantly, compared with that in the WSM3 case, both the MPS heating and CPS heating are all smaller in the inner region near the eyewall (i.e., the region within a radius of 200 km from TC center), but larger in the outer region occupied by outer spiral rainbands (i.e., the region outside a radius of 200 km from TC center) in the WSM6 case. As suggested by Sun *et al.* [2013a], compared with that in the WSM3 case, the less MPS and CPS heating released near the eyewall can lead to less decrease of central pressure and thus smaller pressure gradient near the eyewall in the WSM6 case. Subsequently, under the weaker pressure gradient force, the eyewall in WSM6 case will expand outward, resulting in a much larger size. On the other hand, compared with that in the WSM3 case, there is more MPS and CPS heating released in the outer spiral rainbands in the WSM6 case (Fig. 4.8). As suggested by Wang [2009], MPS and CPS heating in the outer spiral rainbands can make the convection in the outer spiral rainbands more active, which is in favor of the increase of storm size. Thereby, both the less heating near the eyewall and more heating in the outer spiral rainbands all contribute to the increase of the simulated TC size in the WSM6 case, and thus result in the large difference in TC size between WSM6 case and WSM3 case. This could affect the scope of anvil clouds extending over the upper-troposphere over WPSH (Fig. 4.11), and thus contribute to the difference in WPSH intensity and TC track between WSM6 case and WSM3 case (Figs. 4.2 and 4.4), which will be discussed in detail in the following sections.

Phase changes of hydrometeors are important physical processes described in MP. We will investigate the distribution of the hydrometeors and its impact on temperature in the next section.

4.4.2 *Distribution and source of the hydrometeors*

Due to the importance of the phase changes of hydrometeors for the WPSH and TC simulations, it is necessary to compare the hydrometeors distributions of WSM3 and WSM6 experiments with the observations. As mentioned above, we focus on the period before the difference in TC position is insignificant between WSM3 case and WSM6 case. This is because, once the difference in TC position becomes significant, the contribution of MPS may be affected by the difference in the surrounding environment and thus cannot be used to fully explain the difference in hydrometeors distribution between the WSM3 case and WSM6 case.

Fig. 4.8a displays the high resolution (0.1°×0.1°) black body temperature (TBB) observed by FY-2E meteorological satellite, while Figs. 4.8b and 4.8c show two top views of the total mixing ratio of model-simulated hydrometeors for the WSM3 and WSM6 experiments at 0000 UTC 18 October. TBB indicate cloud-top heights if clouds are present below the satellite and are used as an index of convective activity. When equivalent cloud amount and contained hydrometeors increases, TBB decreases, which means deeper convections were observed [Kubota *et al.*, 2005]. Despite the fact that it is meaningless to compare the values of simulated hydrometeors with the values of observed TBB due to their unit difference, we can still use the observed TBB to validate the spatial distributions of the simulated hydrometeors in the WSM3 and WSM6 cases. It is encouraging that both the WSM3 and WSM6 cases well reproduce the main characteristics of the inner and outer spiral rainbands. Although it is not possible to predict the detailed distribution of convective cells along the spiral bands, the model does simulate well the distributions of intense and organized (convective and stratiform) clouds in the spiral rainbands. Both the observed and simulated TC spiral rainbands present strong asymmetry. The inner spiral rainbands are mainly distributed in the east of TC center, while parts of the outer spiral rainbands appear to the southeast, east, and northeast of the TC. On the other hand, compared with that of WSM6 case, the hydrometeors distribution in the WSM3 case is more consistent with the TBB distribution in observation. Both the observed TBB distribution and the simulated hydrometeors distribution in the WSM3 case show that the spiral rainbands are collapsed and break over the aforementioned SR region, which is in the northeast of the TC, while the simulated hydrometeors distribution in WSM6 case displays intense rainbands over the SR region that connects inner and outer spiral rainbands in the northeast of the TC. As mentioned in section 4.1, the difference in the simulated hydrometeors and the related phase change over the SR plays an important role in determining the differences in temperature profile and thus the differences in WPSH intensity between the two cases. We will further discuss this issue in the following.

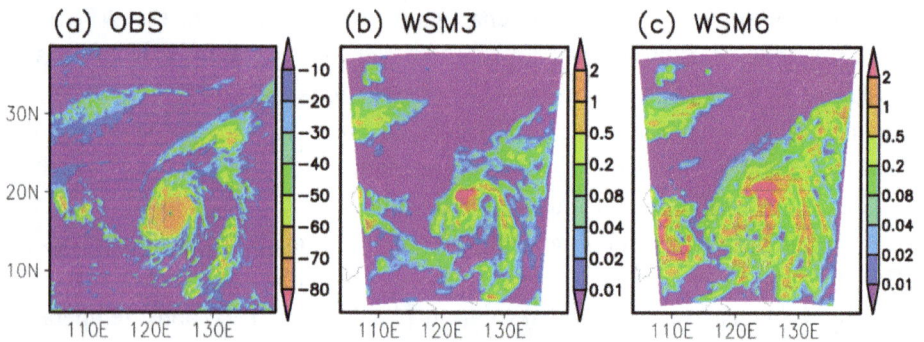

Fig. 4.8. (a) FY-2E black body temperature (TBB; °C) and (b, c) two top views of the total mixing ratio of model-simulated hydrometeors (g kg^{-1}) for WSM3 and WSM6 experiments at 0000 UTC 18 October.

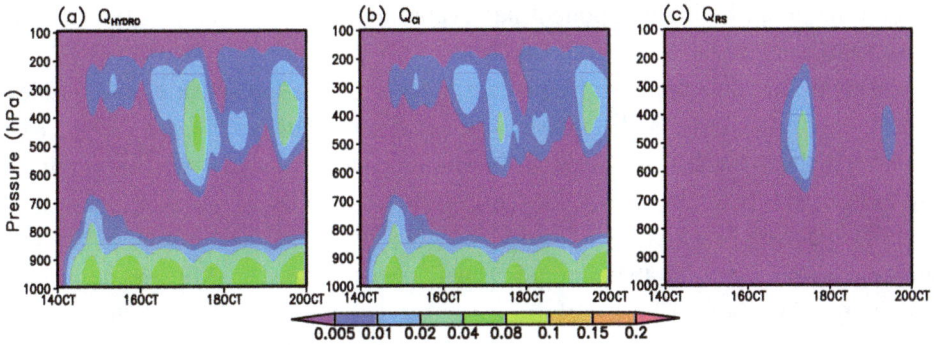

Fig. 4.9. Height-time cross-sections of the simulated hydrometeors and its individual contributions (g kg^{-1}) in WSM3 experiment averaged in SR. (a) total hydrometeors; (b) cloud water/ice; (c) rain/snow.

The WSM3 scheme includes ice and snow processes suitable for mesoscale grid sizes [Hong *et al.*, 2004]. It is a simple ice scheme that predicts water vapor, cloud water/ice, and rain/snow [Skamarock *et al.*, 2008]. It follows Dudhia [1989] to assume cloud water and rain for temperatures below the freezing level, and cloud ice and snow for temperatures above the freezing level. This scheme is computationally efficient for the inclusion of ice processes, and thus widely used in regional climate modeling studies [e.g., Liang *et al.*, 2012; Mooney *et al.*, 2013]. Note that not every hydrometeor is stored as individual array and output from the model in the WSM3 scheme. Cloud water and ice are stored as the cloud component of the hydrometeors in the output from the model, while rain and snow are stored as the precipitating component [Braun, 2006]. Fig. 4.9 shows the height-time cross-sections of the simulated hydrometeors averaged over the SR in WSM3 experiment. The simulated hydrometeors are mostly distributed in the levels between 600-200 hPa and below 850 hPa. Consistent with the microphysical heating, the hydrometeors in the upper levels become notable after 0000 UTC 17 October when the simulated storm is approaching SR (Fig. 4.9a). The distribution of the cloud component (i.e., cloud water and ice) is basically similar to that of total hydrometeors, especially in the levels below 850 hPa (Fig. 4.9b). As suggested by Dudhia [1989], the cloud component in the low (middle and upper) troposphere below (above) freezing level can be taken as cloud water (ice). The magnitude and vertical extent of the cloud component are all much larger than that of precipitating component (Fig. 4.9c). Thus, the cloud component can be considered as the major component of the total hydrometeors, while the precipitating component is minor. In addition, the precipitating component is only distributed in the levels between 650-300 hPa in the form of snow. From 0000 UTC 17 to 1200 UTC 17 October, no precipitation occurs at the surface, implying that all the precipitating component has evaporated before arriving at the surface (Fig. 4.9c).

The WSM6 scheme is similar to the WSM3 simple-ice scheme but includes descriptions of graupel and its associated processes. Different from the WSM3 scheme, in the WSM6, water vapor, rain, snow, cloud ice, cloud water, and graupel are held in six different arrays. Hence it allows supercooled water to exist, and describes the gradual melting of snow and graupel during their falling below the melting layer. Some of the graupel-related terms in the WSM6 are described following Lin *et al.* [1983], but its ice-phase behavior is different due to the changes of Hong *et al.* [2004]. Fig. 4.10 shows the height-time cross-sections of the simulated hydrometeors in the WSM6 experiment averaged in SR. Compared with the results in the WSM3 experiment, the WSM6 experiment generates more than twice the value of area-averaged total hydrometeors in the WSM3 experiment. Moreover, the total hydrometeors in the WSM6 experiment are mostly concentrated in the middle troposphere after 0000 UTC 17 October, with a maximum center at about 450 hPa (Fig. 4.10a). The total hydrometeors in the WSM6 case are divided into cloud (i.e., cloud water and ice) and precipitating components (i.e., rain, snow, and graupel). The cloud water is concentrated in the low troposphere below 900 hPa (Fig. 4.10b). The cloud ice is concentrated over the levels between 600-200 hPa and about two times larger than that appearing above freezing level in the WSM3 (Fig. 4.10c). A small amount of rain water, which is absent in the WSM3 case, appears in the levels below 600 hPa after 0000 UTC 17 October (Fig. 4.10d). Compared with the other two precipitating hydrometeors (i.e. rain and graupel), the simulated snow exhibits a much larger value and vertical extent, and thus can be considered as the major part of the precipitating component. Consistent with that in WSM3, snow is concentrated in the levels between 600-200 hPa after 0000 UTC 17 October in the WSM6 case (Fig. 4.10e). The magnitude of snow in the WSM6, however, is about three times larger than that in the WSM3 and contributes greatly to the difference in hydrometeors between the WSM6 case and WSM3 case. Moreover, the WSM6 case with the higher concentration of snow produces more stratiform precipitation compared to the WSM3 case (Figs. 4.10d and 4.10e). This result is consistent with other modeling studies [Lord *et al.*, 1984; McCumber *et al.*, 1991]. The simulated graupel only exists in a limited area near 550 hPa, and its value is much smaller than that of rain in the WSM6 case (Fig. 4.10f).

Overall, both cloud and precipitating components in the WSM6 show much larger values and vertical extents than those in the WSM3 case. This is an important reason for the difference in DH and temperature between the WSM6 and WSM3 cases. Since there is low water condensation in the WPSH region [Noone *et al.*, 2011; Noone, 2012], the large amount of the simulated hydrometeors in the WSM6 over the SR must be transported from somewhere else.

To understand the source of the hydrometeors in the SR, we depict the simulated hydrometeors at different levels (e.g., 300 hPa, 500 hPa, and 700 hPa) for the WSM3 and WSM6 cases in Fig. 4.11. Compared to that in the WSM3, the simulated hydrometeors in the WSM6 extend to a much larger area far away from the storm center in the upper-troposphere despite of the less amount of hydrometeors near the eyewall.

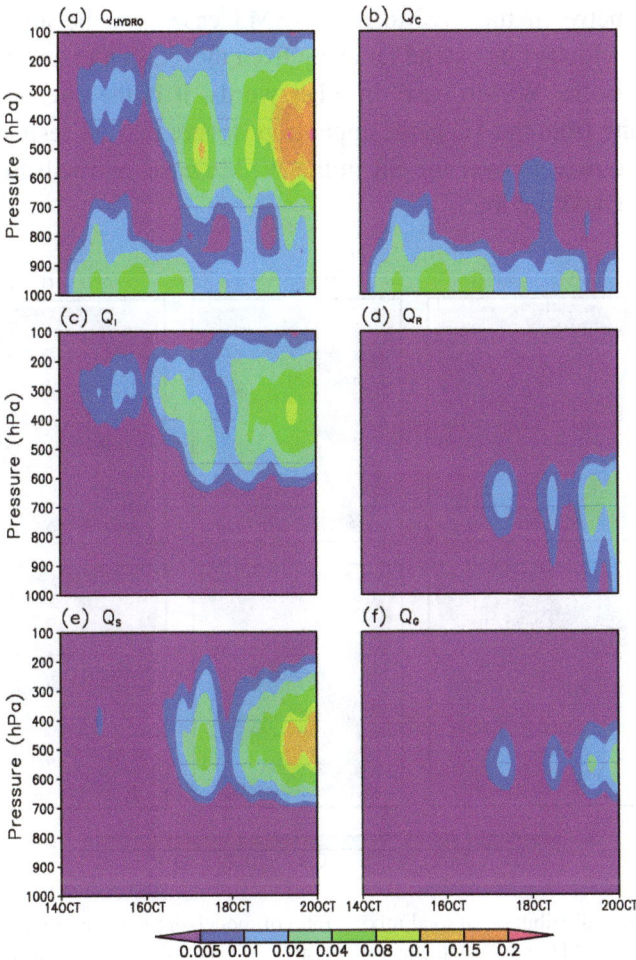

Fig. 4.10. As in Fig. 4.9, but for the simulated hydrometeors and its individual contributions (g kg^{-1}) in WSM6 experiment: (a) total hydrometeors; (b) cloud water; (c) cloud ice; (d) rain water; (e) snow; (f) graupel.

Note that, the difference in hydrometeors between the WSM6 and WSM3 also varies with height. The WSM6 produces much larger amounts of anvil clouds above 500 hPa in terms of hydrometeors in the SR. These anvil clouds are far away from the storm center and can reach the SR, which is consistent with the large amount of cloud ice and snow in Figs. 4.10c and 10e. At 700 hPa level, the WSM6 produces more hydrometeors in terms of precipitation in the SR, which is consistent with the larger value of rain water in Fig. 4.10d. However, there is no significant difference in hydrometeors between the two cases at 0000 UTC 15 October before the storm entering into the model domain. The difference in hydrometeors between the WSM6 and WSM3 cases can be attributed to the excessive anvil clouds that extend far away from the TC center and reach the area of the WPSH in the WSM6 case. In other words, it is the overestimated anvil clouds extending from the TC in the upper troposphere that are responsible for the excessive hydrometeors over the SR in the WSM6 case. We will further discuss this issue in the next section.

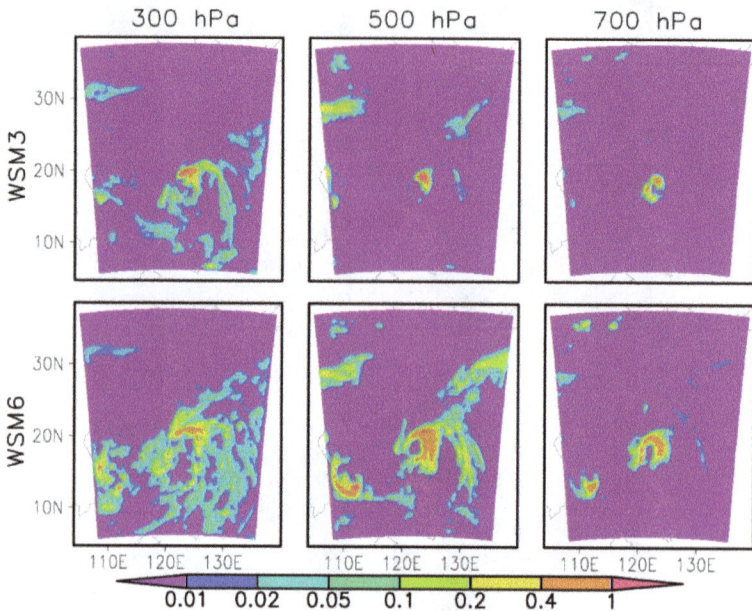

Fig. 4.11. Horizontal distribution of total mixing ratio of model-simulated hydrometeors (g kg⁻¹) at different levels for WSM3 and WSM6 experiments at 0000 UTC 18 October, respectively.

4.4.3 *The role of phase change*

As discussed above, the microphysical heating (i.e., DH) caused by phase changes of water substance and excessive anvil clouds in the WSM6 are responsible for the temperature difference between the WSM6 and WSM3 cases. Thus, it is necessary to investigate the hydrometeors and related phase changes in TC region as the storm is approaching the SR.

Fig. 4.12 shows azimuthal-averaged cross-sections of microphysical heating and temperature in the two experiments and their difference at 0000 UTC 18 October. The grid spacing of 20 km is too coarse to well reproduce the structure of TC eyewall [Fierro *et al.*, 2009; Sun *et al.*, 2013a], and the radius of the simulated eyewall in the two cases is notably larger than the observed radius. The microphysical heating is mostly concentrated in the area of eyewall convections, and extends outward with height following the eyewall slope. Stern and Nolan [2009] showed that the outward slope of the eyewall with height is directly proportional to the radius of the eyewall. As expected, the WSM6 case is characterized by a less upright (i.e., larger slope) eyewall in terms of the latent heat release due to the larger eyewall radius when compared with the WSM3 case. As a result, underneath the more tilted eyewall in the WSM6 case, the downdrafts induced by evaporation of rain and melting of snow and graupel make the sub-cloud layers inflow drier and cooler than that in WSM3, resulting in lower temperature below 500 hPa and in boundary layer (Figs. 4.12a and 12b). This result is consistent with modeling studies of Yang *et al.* [2007]. Comparing with results in the WSM3 case, the WSM6 case can produce more microphysical latent heating near the eyewall on one hand, and exhibit a feature of cooling in the upper levels and heating in the lower levels which can promote the development of convection by increasing the convective instability on the other hand. This leads to a stronger secondary circulation, and support a larger low-level inward mass flux and larger volumes of mixed-phase particles aloft near the eyewall in the WSM6 case than in the WSM3 case.

As suggested by Powell [1990], the stratiform precipitation of the storm extends outward as anvil rain, and some of the strong anvil showers in the WSM6 case can generate penetrative downdrafts below the anvil cloud. This leads to an extended heat source due to condensation in the anvil cloud and an extended heat sink due to precipitation evaporation below the anvil cloud [Willoughby, 1988]. Note that, the anvil clouds is being used to refer to cloud water and cloud ice, while the anvil showers is being used to refer to rain, snow and graupel in this study. Because of the wider and thicker anvil cloud in the WSM6 case, both the warming above and the cooling below 500 hPa are all stronger than that in the WSM3 case.

This mechanism explains the temperature difference in outer region of the TC and SR (Figs. 4.12c and 4.5b). Consistent with Stossmeister and Barnes [1992], the changed temperature profile would cause decreasing of geopotential height at 500 hPa in SR (Fig. 4.5a) and weakening of WPSH (Fig. 4.4i), leading to an earlier northward-turning of TC in the WSM6 case compared to that in the WSM3 case (Fig. 4.2).

To further investigate the impacts of MPS on temperature distribution, we examine individual component (i.e., cloud and precipitating component) related to microphysical heating. Fig. 4.13 shows the azimuthal-averaged cross-sections of cloud and precipitating components and hydrometeors in the two cases and their differences at 0000 UTC 18 October. As suggested by Wang [2002], cloud water is formed by condensation of supersaturated water vapor; thus, a high concentration of cloud component is closely related to the outward-tilted updrafts in the eyewall (Figs. 4.13a and 4.13b). In the WSM3 case with simple-ice phase, the cloud component (i.e., cloud water and ice) is concentrated in the eyewall region with a maximum at about 600 hPa. It contributes greatly to the eyewall updraft due to the related condensation warming effects (Fig. 4.13a). The distribution of cloud component in WSM6 case is similar to that in WSM3 case, except for a larger value and broader scope (Fig. 4.13b). This induces an excessive heating not only in the storm eyewall region, but also in the outer region of the storm especially in the levels above 500 hPa (Fig. 4.13c). This is the main reason for higher temperature above 500 hPa in the WSM6 case (Fig. 4.12c).

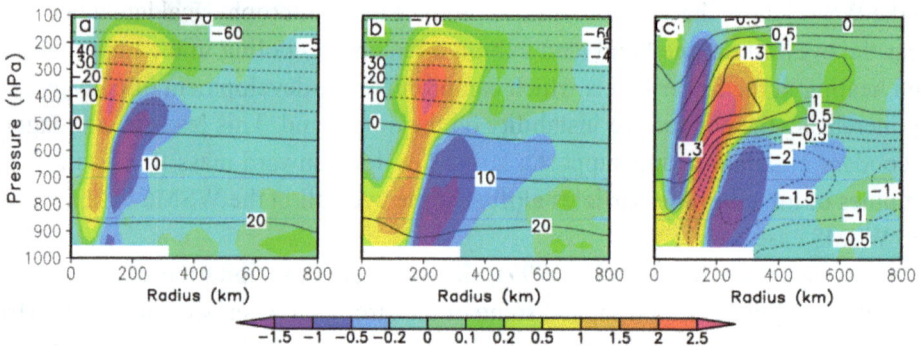

Fig. 4.12. Height-radius plots of mean azimuthal microphysical heating (°C h^{-1}; shaded) and temperature (°C; contoured) for the WSM3 case (a), the WSM6 case (b), and the difference between WSM6 case and WSM3 case (c) at 0000 UTC 18 October.

The precipitating component in both WSM3 and WSM6 cases is mostly concentrated in the middle- and upper-troposphere (above freezing level) in the form of snow, which melts into rain in the saturated region or sublimates to water vapor in the unsaturated region as it falls through the freezing (or melting) level. Below the freezing level, rainwater forms through conversion from the cloud water, and grows by collecting cloud water and by melting of both snow and graupel, resulting in a large concentration there. Note that along with the melting of snow, the evaporating of rain during its falling process is responsible for the cooling in the low troposphere below the freezing level (Figs. 4.13d and 4.13e). Although the maximum value of precipitating component in the WSM6 case is not larger and sometimes smaller than that in the WSM3 case, its extent is larger in terms of hydrometeors due to the larger eyewall slope. This results in a much greater extent of cooling below the freezing level due to melting and evaporating effects (Fig. 4.13f), which contributes greatly to the lower temperature below 500 hPa in the outer region of the storm in the WSM6 case (Fig. 4.12c) when compared with that in the WSM3 case.

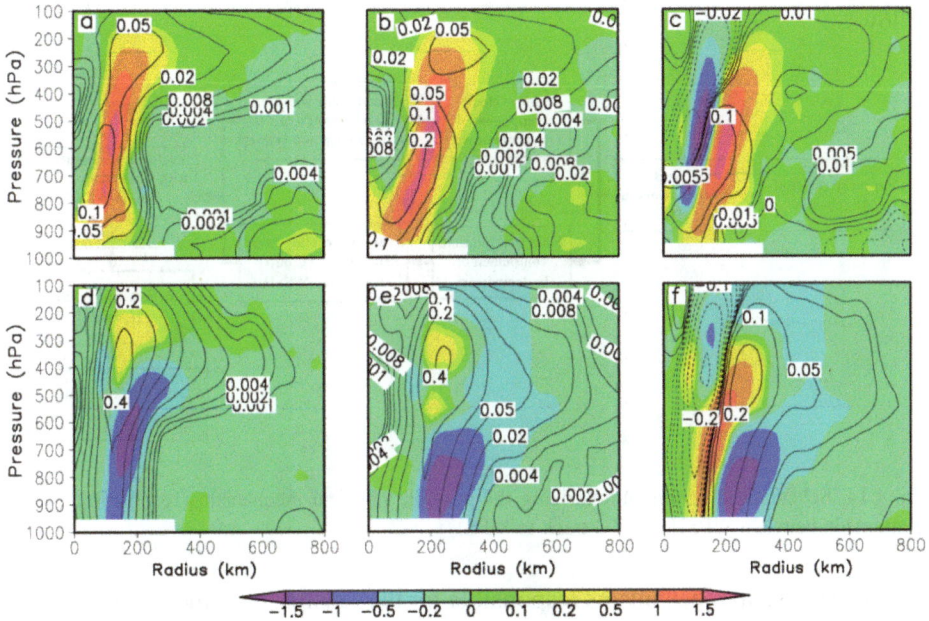

Fig. 4.13. Height-radius plots of mean azimuthal cloud (top) and precipitating (bottom) components of microphysical heating (°C h^{-1}; shaded) and hydrometeors (g kg^{-1}; contoured) for the WSM3 case (a, d), the WSM6 case (b, e), and the difference between WSM6 case and WSM3 case (c, f) at 0000 UTC 18 October.

Overall, compared with the WSM3 case, it is the difference in cloud (precipitation) component that is responsible for the higher (lower) temperature above (below) the freezing level in the outer region of the storm in the WSM6 case.

As the storm approaches SR, the outer region of storm extends and reaches the SR at 0000 UTC 18 October. The temperature profile in the SR is basically consistent with the temperature profile in the outer region of the storm and leads to a decrease in the geopotential height at 500 hPa, which subsequently contributes to the unrealistic break of WPSH in the WSM6 case.

4.4.4 *Possible mechanisms*

As mentioned in the section of PVT diagnosis, the failure in simulation of the TC motion (e.g. turning ahead of schedule) is attributed to the unrealistic break of WPSH in SERs. Further analysis reveals that the large bias in DH over the area of TC convections is responsible for the break of WPSH in the SER when the center of SR is far away from the TC center. How the MPS affects the DH and results in the break of WPSH and an earlier turning of TC in the SERs is summarized as follows.

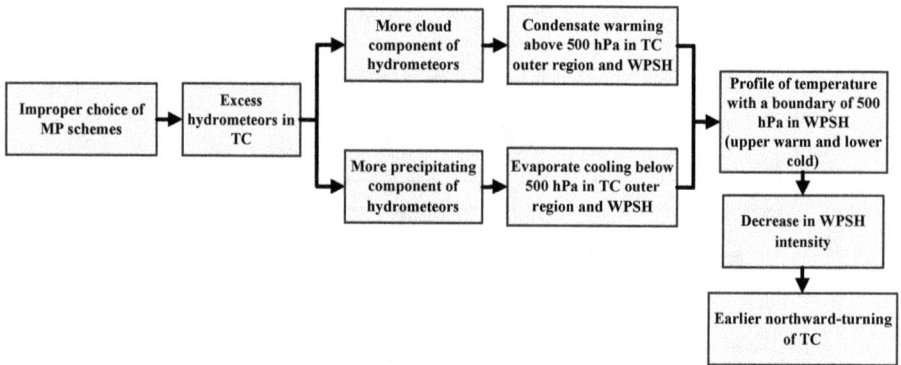

Fig. 4.14. Schematic diagram summarizing the possible physical mechanisms responsible for the relationship between WPSH and TC track to MPS.

The physical mechanisms responsible for the sensitivity of the WPSH and TC track to the MPSs are identified through comprehensive diagnosis and are schematically summarized in Fig. 4.14. Compared with the CTR, the simulations with different MPSs in SERs overestimate hydrometeors in the storm eyewall region. The excess hydrometeors in SERs extend outward with height following the eyewall slope and reach the outer region of the TC in upper-troposphere, and thus result in excessive anvil clouds and showers (i.e., cloud and precipitating components) in the upper- and mid-troposphere. As the simulated TC approaches the WPSH, the excessive anvil clouds in SERs extend far away from the TC center and reach the area of WPSH. The overestimated cloud and precipitating components in SERs cause condensation warming above 500 hPa and evaporative (melting) cooling below 500 hPa (freezing level) in the outer region of the TC. Such a temperature pattern is quite clear over the SR when the TC is approaching, leading to a decrease of 500 hPa geopotential height and the unrealistic break of WPSH in SR. This in turn prompts the earlier turning of TC.

4.5 Discussions

For the sensitivity of TC motion and WPSH to MPSs, it is found that the model simulation of WPSH is usually realistic under normal conditions with the absence of TC activities, but it is quite sensitive to the MPS when TCs are active over the WNP. Unrealistic representation of microphysical process in TCs is responsible for the failed simulation of WPSH. In this study, the simulation using the WSM3 scheme presents a much better result in terms of TC track and WPSH intensity, compared to results using the other three MPS schemes. Consistent with the previous studies, the TC motion is sensitive to MPS schemes.

It is noteworthy there is no MPS appropriate for all TC cases due to the specificity of each TC case. Although the simulation with WSM3 scheme produces a better result than those with the other three MPSs in this study, it does not necessarily mean the WSM3 scheme is more suitable than other MPs for the simulation of WPSH when TCs are active over the WNP. The focus of this paper is on the sensitivity of the simulated TC motion and WPSH to MPS schemes, instead of on the advantage of WSM3 scheme in simulating the WPSH. Our conclusions are based on the comparisons between sensitivity experiments and less dependent on the simulation results.

In this study, the simulated storm intensities are not presented because the 20-km grid spacing is too coarse to simulate the storm intensity accurately. For this reason, numerous RCM studies with coarse resolution are conducted to investigate the TC track and frequency rather than the TC intensity. On the other hand, the TC intensity may not be important in our efforts to understand how the TC feedback affects the simulation of WPSH intensity. Although the simulated TC intensity in SERs is weaker than that in the CTR during the track turning, it is the TC in SERs rather than that in CTR that weakens the simulated WPSH intensity, resulting in the unrealistic break of WPSH and subsequently the early turning of the TC. Thereby, the TC intensity plays an insignificant role in determining the simulation of WPSH intensity and TC track. Hence, the TC intensity analysis doesn't make much sense in the present study.

The potential vorticity tendency (PVT) diagnosis technique is utilized to estimate the contributions of the horizontal advection (HA), vertical transportation

(VT) and diabatic heating (DH) to TC motion. Their differences between the SER and CTR simulations are analyzed. HA makes the largest contribution to the difference in TC motion between the SER and CTR. It slows down the westward moving but accelerates the northward moving of TC in the SER. VT and DH also affect the TC motion but their impacts are less significant than that of HA. It is the difference in HA that is responsible for the different TC motion simulated by the SER and CTR, especially Megi's unrealistic early northward-turning in SER. The MPS directly contributes to the early turning of TC in SERs by changing the environmental flow (i.e., steering flow) of the TC. In other words, due to the unrealistic break of WPSH, the steering flow in the south of WPSH is changed, and results in the early turning of TC in SER.

Further analysis shows that the TC feedback on the intensity of WPSH plays an important role in determining the break of WPSH and thus the early turning of TC in SERs. The physical mechanisms responsible for the sensitivity of the WPSH and TC track simulation to MPSs are identified through comprehensive diagnostic analysis. The large bias in DH simulation over the area of TC convections is responsible for the break of WPSH in SERs. Specifically, compared with the WSM3 in CTR, the failure in the simulation of the WPSH intensity and TC track using the other MPSs in SERs is attributed to the overestimation of anvil clouds, which extend far away from the TC center and reach the area of the WPSH. On the other hand, compared with that in the CTR, both the less heating near the eyewall and more heating in the outer spiral rainbands in SERs all contribute to the increase of the simulated TC size, and thus result in the large difference in TC size between SERs and CTR. This contributes to the difference in the scope of anvil clouds extending over the upper-troposphere over WPSH, and thus the difference in WPSH intensity and TC track between SERs and CTR. The MPSs used in SERs produce excessive hydrometeors in the outer region of the TC with more cloud and precipitation, which are reflected in the excessive anvil clouds and showers. This leads to a condensation warming above 500 hPa and an evaporative cooling below 500 hPa (freezing level) in the TC outer region, which results in the large bias in the simulations of DH and thus temperature profile. As the TC is approaching the WPSH, such a pattern of vertical temperature distribution causes the decrease of 500 hPa geopotential height and subsequent break of WPSH. Errors in the WPSH simulation described above change the large-scale steering flow, leading to the early turning of TC. Through a series of feedback loop the model eventually fails in the simulation of both TC and WPSH. It is important to note that, although the simulated intensity of the TC is much weaker than the observation, it does not imply that the "feedback" of the simulated TC to the simulated WPSH is insignificant. MPSs cannot directly impact the simulation of

WPSH without the feedback of TC, however, it may indirectly impact the simulation by affecting the TC activity. In another word, the MPS affects the simulation of the TC activity (e.g., eyewall convections, anvil cloud and showers, hydrometeors distributions, etc.), which subsequently influences the simulation of WPSH intensity. The TC feedback on the intensity of WPSH plays a critical role in the model behavior of simulation for both the WPSH and TC.

Similar results are found if we compare the WSM3 case and the other two cases, i.e. Lin and Thompson cases. This is because vapor, rain, snow, cloud ice, cloud water, and graupel are held in six different arrays in the all three MPSs (i.e. WSM6, Lin, and Thompson) used in SERs. Different from that in the WSM3 scheme, supercooled water is allowed to exist in the three schemes. Gradual melting of snow and graupel falling below the melting layer are also described in the three MPSs used in SERs. As a result, all the three SERs produce a similar volume of hydrometeors in the outer region of the TC, which is much larger than that in the WSM3 case especially in terms of cloud ice and snow. Note that the three experiments in SERs have the similar simulation results in terms of the TC motion, WPSH intensity, and vertical distribution of hydrometeors and temperature. In fact, compared to WSM3 scheme, the increase in the category of hydrometeors in the other three MPSs may improve the realism of the results. However, just as Mass *et al.* [2002] suggested, it does not necessarily ensure the improvement of the forecasts. This is because the increase in the category of hydrometeors may introduce extra errors in the simulation. In our present study, all the three MPs in the SER experiments overestimate the hydrometeors in the TC, resulting in large biases in simulating the WPSH and TC track. It is necessary to improve the available MPs or develop a new MPS that is appropriate for the TC-WPSH simulations. This will be a topic for our future research. In addition, although the CTR with the WSM3 scheme produces the best results in this study, it does not necessarily imply that simulations with the WSM3 scheme can always perform better than other MPs for different TCs and WPSH simulations since the performance of a specific physics parameterization scheme may be case dependent.

Results of this study are helpful to understand the impact of various MPSs on the WPSH and TC track simulation. The physical mechanism revealed in this study not only helps us better understand the TC and WPSH dynamics, but also provide insightful knowledge for further improvement in the MPs. It is noteworthy that, in addition to the MPS scheme, the model-simulated anvil clouds may also be sensitive to other model physics schemes (e.g. cumulus, boundary layer parameterization schemes, etc). In the future work, we will investigate how changes in these physics schemes affect the model simulation of WPSH and TC track.

Chapter 5

Mechanism of Tropical Cyclone Track Sensitive to Planetary Boundary Layer Scheme

5.1 Introduction

With the progress of numerical weather prediction (NWP) and operational forecast, errors in the prediction of TC track have been significantly reduced. However, great difficulties still exist for realistic track simulation/prediction of those TCs that experienced abrupt changes in their moving paths. One major reason for such difficulties is that the understanding of the mechanisms for the abrupt change in TC movement is still very limited [Ni *et al.*, 2013]. Many factors such as the large scale steering flow, the mesoscale and microscale convective systems in the surrounding area of the TC, and effects of complex terrain etc. can affect the movement of TCs and lead to abrupt changes. Yu *et al.* [2012] found that the primary weather systems are different for TCs at various regions. However, in-depth studies of the mechanisms for abnormal and abrupt changes in TC movement caused by the interaction between TC and the large-scale circulation system are very limited.

For those TCs that affect China, their tracks are largely dependent on the interaction between the WPSH and TC [Zhong, 2006]. Wang *et al.* [1991] found that the subtropical high is the large scale circulation system that directly affects TCs, which either stay offshore and move up northward or experience abrupt turning in the remote oceans. Wei *et al.* [2010] analyzed the track of typhoon Sinlaku (2008) and found that the intensification and westward extension of the subtropical high was the major reason for Sinlaku to turn westward. Meanwhile, TC feedback could influence the intensity and influencing area of the subtropical high. Li *et al.* [2002] indicated that the interaction between TC and the subtropical high could result in intensification and westward extension of the subtropical high. Sun *et al.* [2014a, 2015a] revealed that once the ice crystals and other hydrometers in the upper wall clouds of TC dispersed to the subtropical high region, the latent heat release and absorption related to phase changes of these hydrometers could weaken the subtropical high and eventually led to changes in TC track. The above studies indicate that it is necessary to investigate the interaction and feedback between the subtropical high and TC, particularly the feedback of TC on the subtropical high, which will be helpful for further exploration of the mechanisms for TC movement and TC track change.

Numerical modeling study is an important approach for TC simulation and TC track forecast [Zhang *et al.*, 2007; Han *et al.*, 2008; Guan *et al.*, 2011]. Various physical parameterization schemes in numerical models have great impacts on the simulation of TC development and movement. The PBL scheme is one of those physical schemes that have large impacts on TC simulation. A previous study [Li *et al.*, 2005] revealed that the vertical transport of sensible and latent heat fluxes and atmospheric moisture is critical for the genesis and development of TC. The PBL schemes in various models are developed based on different principles, and thus their descriptions of PBL in the TC are different, which could lead to distinct differences in TC simulation and forecast. At present, some studies have been conducted to investigate impacts of various PBL schemes on the simulation of TC structure and intensity. Large differences were found in the minimum sea level pressure (MSLP) and wind speeds of hurricane Bob simulated by MM5 using different PBL schemes [Braun *et al.*, 2000]. Simulations of typhoon Dan using different PBL schemes indicated that differences in the PBL simulation could lead to changes in the typhoon size and the simulated horizontal and vertical circulations were also different [Deng *et al.*, 2005]. Nolan *et al.* [2009] applied WRF to simulate Hurricane Isabel (2003) using two different PBL schemes and compared the simulated maximum wind speed and PBL structure in the periphery of Isabel. Lai *et al.* [2010] compared the simulations of Typhoon Lolave (2009) with different PBL schemes and found that the YSU PBL scheme yielded a more realistic result compared to other PBL schemes. Zhou *et al.* [2013] explored impacts of various PBL schemes on the track and intensity simulation of Typhoon Megi (2010) and found large differences in the simulated water vapor flux in PBL, which resulted in differences in the simulated TC intensity. Although the above studies have shown that the PBL scheme has great impacts on the simulation of TCs, the mechanisms behind the PBL impacts still remain unclear. In particular, little is known about the PBL impact on the interaction between the subtropical high and TC.

In this Chapter, the WRF model is applied to simulate Megi (2010) again, with a focus on the mechanism of PBL scheme influence on TC track. The relationship between TC size and the subtropical high intensity is investigated to reveal the possible mechanisms for PBL scheme impacts on TC track simulation.

5.2 Experimental design

The data used include the NCEP global $1° \times 1°$ analysis data (FNL), the 3-hourly real-time dataset on global $0.25 \times 0.25°$ grids of Tropical Rainfall Measuring Mission (TRMM) (hereafter T-3B42), a joint mission of NASA and the Japan Aerospace Exploration Agency, and the best track data at 6-hour intervals published by JTWC.

Same as the simulations for the sensitivity of Megi track to CPS in Chapter 3 and MPS in Chapter 4, the model domain was centered at (122°E, 22°N) with horizontal grid numbers of 160180 at the grid interval of 20km. There were 36 σ levels in the vertical. Physical parameterization schemes included the WSM 3-class microphysics scheme [Hong *et al.*, 2004], the GD cumulus scheme [Grell and Dévényi, 2002], the RRTM longwave radiation scheme [Mlawer *et al.*, 1997], and the Goddard shortwave radiation scheme [Chou and Suarez, 1994]. Two sets of surface and PBL schemes, i.e. the MM5 surface layer scheme [Beliaars, 1995] coupled with the YSU PBL scheme, and the Eta surface layer scheme coupled with the MYJ PBL scheme [Janjić, 2002], were used in the simulation. The thermal diffusion scheme scheme [Skamarock *et al.*, 2005] was applied to simulate soil physics. The initial condition and lateral boundary forcing were extracted from the global $1°\times1°$ NCEP/NCAR reanalysis product. The model was integrated for 264 hours, covering the period from 0000 UTC 14 October to 0000 UTC 25 October 2010. The entire lifespan of Megi from its genesis and development to landing were included in the integration period. The time step was 90s.

WRFV3.3 includes 11 local and nonlocal PBL schemes. In the Chapter, the PBL scheme of local Mellor-Yamada-Janjic (MYJ) and nonlocal Yonsei University (YSU) [Noh *et al.*, 2003] were used to simulate the TC track respectively, and the difference in the TC track simulation by the two PBL schemes was analyzed to explore the mechanisms for the difference. In the MYJ scheme, turbulent fluxes at each model grid are determined by the averages of physical variables at the grid; in the YSU scheme, turbulent fluxes at each individual grid are determined by physical variables at the grid and its surrounding grids [Zhang *et al.*, 2012]. The MYJ scheme is a PBL scheme with local treatment of large-scale eddies, it is developed to describe turbulent motion above the surface layer based on the Mellor-Yamada 2.5-order turbulence closure scheme. Local turbulent momentum is used in the MYJ to represent the vertical mixing in the PBL and free atmosphere above. Turbulent diffusion coefficients are determined based on the turbulent kinetic energy prediction equation in the MYJ scheme. The turbulent kinetic energy and its dissipation rate are obtained through iterative method, while the turbulence length scale is also modified to make it appropriate for high resolution PBL simulation. The MYJ scheme performs well for stable and weakly unstable PBL, but it induces large biases in convective PBL simulation.

The non-local YSU scheme is a first-order closure scheme based on the K-theory. A counter-gradient term is added to the turbulence diffusion equation to consider the entrainment process of heat and momentum fluxes at the inversion layer, which effectively increases the mixing layer height and avoids the problem of treating turbulence as mixing within the PBL. The Bulk Richardson number is used in the YSU scheme to determine the PBL height, and the influence of temperature on the PBL height is also considered. The counter-gradient term is added to the vertical profile for turbulent transport and the non-local k-mixing for momentum is considered.

5.3 Simulation results with different PBL schemes

5.3.1 *TC track*

The track of Megi from the JTWC best track data and from the WRF simulations using the two different PBL schemes was displayed in Fig. 5.1, which shows that the simulated TC using both the YSU and MYJ schemes moved westward steadily following almost the same track consistent with the observation before Megi made landfall in the Philippines. However, large differences appeared in the simulations with these two schemes before Megi moved across the northern Philippines and entered the SCS. In the simulation with the YSU scheme, Megi turned northward earlier than observation after 1200 UTC 18 October, passed by eastern Taiwan and landed in the coastal region of Zhejiang province. The simulated track is quite different to the observation. However, in the simulation with the MYJ scheme, Megi continued its westward moving after 1200 UTC 18 October and entered the SCS. One day later, it turned northward abruptly. The simulated TC track is highly consistent with the observation. The above results suggest that the TC track simulation is sensitive to PBL scheme.

5.3.2 *Geopotential height at 500 hPa*

Zhu *et al.* [2000] proposed that changes in the WPSH have great impacts on TC track. The moving of Megi was largely determined by the large-scale steering flow. Fig. 5.2 presents the geopotential height at 500 hPa from reanalysis product and from simulations using the two PBL schemes. At 1200 UTC 14 October, Megi was located to the east of the model domain. The comparison of simulations at this time showed that the WPSH was zonally distributed with similar intensity and location in the two experiments using YSU and MYJ PBL schemes respectively (Fig. 5.2a and 5.2b), and the simulation results are consistent with FNL analysis (Fig. 5.2c). The above results correspond to the situation when no active TCs are found within the model domain. In this situation, the simulated large-scale circulation and WPSH are not sensitive to the PBL scheme. By 0000 UTC 18 October, Megi reached the ocean to the northeast of the Philippines (Fig. 5.2f). At this time, although the simulated TC center was similar in the two simulations, large difference appeared in the simulated size and intensity of both Megi and the WPSH.

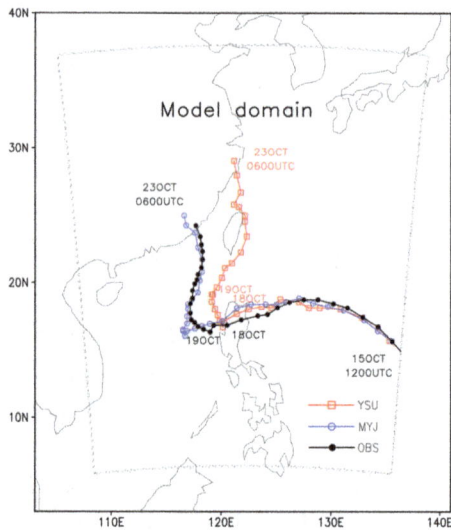

Fig. 5.1. The model domain (sector area) and the track of Meg (2010) from observation and simulations.

The TC size in the simulation using the YSU scheme was relatively large, and the ridgeline of the WPSH reached the Taiwan Strait (Fig. 5.2d). In contrast, the TC size simulated by the MYJ scheme was relatively small (Fig. 5.2e), while the simulated WPSH extended westward to the coastal region of southern China and remained almost intact. Compared with the FNL analysis (Fig. 5.2f), it was found that the TC size was overestimated while the WPSH broke earlier than normal and shifted eastward by using the YSU scheme; in contrast, the TC size and the WPSH intensity and location were close to FNL in the simulation using the MYJ scheme. The above results suggest that the PBL scheme would not directly affect the simulation of the WPSH intensity, however, it indirectly affects the WPSH intensity via affecting the TC size.

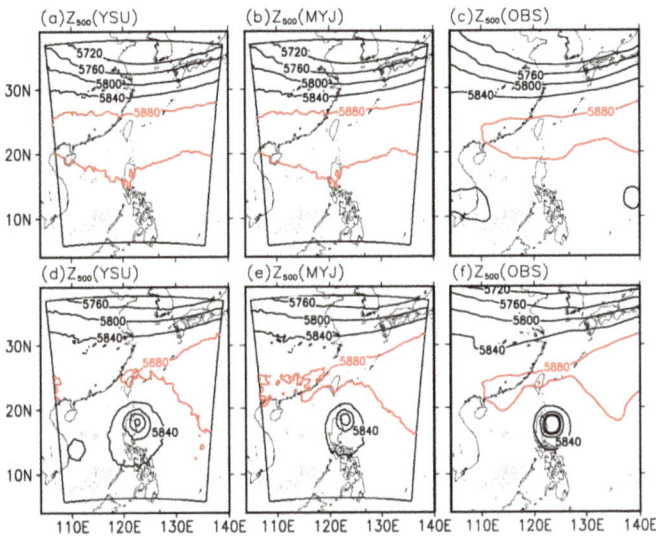

Fig. 5.2. Observed and simulated geopotential height (unit: gpm) at 500 hPa at 1200UTC 14 Oct. (a, b, c) and 0000UTC 18 Oct (d, e, f). (a, d) Simulations using the YSU scheme; (b, e) simulations using the MYJ scheme; (c, f): FNL reanalysis.

5.3.3 *TC size*

In this study, the area enclosed by the closed 1000 hPa isobar in the sea level pressure chart (ACI) was defined as the TC size index. Time series of ACI simulations using the two PBL schemes were displayed in Fig. 5.3, indicating that the temporal evolution of ACI in the two simulations basically was similar to the observation, which increased first and then decreased with time. However, the ACI in the simulation with the YSU scheme was always larger than that simulated with the MYJ scheme. Relatively speaking, the TC size simulated by using the MYJ scheme is more consistent with the observation. In addition, the TC size simulated by the MYJ and YSU schemes was similar before Megi made landfall in the Philippines, but the TC size simulated by the YSU scheme rapidly increased and the difference in the TC size simulated by the two PBL schemes gradually increased since 1200 UTC 16 October, shortly after Megi moved into the model domain. The difference in the simulated TC size between the two PBL schemes reached the maximum during 0000 UTC 18 -0000 UTC 19 October (the period when Megi turned earlier than observation in the simulation with the YSU scheme). As Megi was approaching the land, the TC size simulated with the MYJ scheme and from the observation both rapidly decreased, but it decreased slowly in the simulation with the YSU scheme.

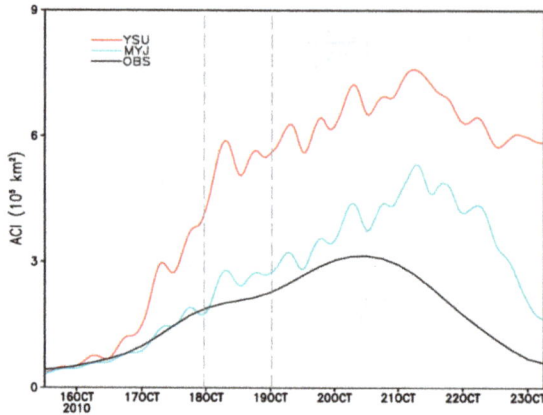

Fig. 5.3. Observed (OBS) and simulated temporal evolution of TC size (ACI) with YSU scheme and MYJ scheme respectively (unit: 10^5 km²; the dashed line represents the period when the simulated TC track deviation with two schemes appears before turning).

Sun *et al.* [2015b] indicated that the pressure gradient in the peripheral region of TC increased with the TC size. Gopalakrishnan [2011] proposed that the inward radial velocity would increase with the increase in the pressure gradient in the peripheral region of TC, leading to larger inflow mass flux (IMF) from the peripheral region to the TC center. Thereby, the larger the simulated TC size is, the larger the pressure gradient will be in the peripheral region of TC, and the larger the IMF will be. IMF can be expressed by:

$$\text{IMF} = \int_0^{1000} \int_0^{2\pi} \int_0^R u_r(z,\theta,r)\ \rho(z,\theta,r)\ dz d\theta dr \tag{5.1}$$

where u_r, ρ are the radial velocity and air density, respectively; r is the distance to the TC center, z is the altitude above the sea level, θ is the azimuth with the center located at the TC center, R is set to 800km.

The simulated IMFs below 1km within the radius of 800km of the TC center were presented in Fig. 5.4, which shows that the TC size was similar in the two simulations with different PBL schemes before 1200 UTC 16 October, but the simulated IMFs were already different at this time. The simulated IMF with the YSU scheme was obviously larger than that with the MYJ scheme, and this phenomenon persisted in the subsequent time. This indicates that with the increase in TC size, more air mass were transported to central TC area in the simulation with the YSU scheme. Furthermore, the inflow mass flux transported to the central TC area demonstrated a distinct diurnal cycle. As a deep low-pressure system, when the TC was approaching a high pressure system such as the WPSH, large pressure gradient existed between the WPSH and the TC. As a result, huge amounts of IMF were transported from the WPSH region to the TC region, leading to a weakened WPSH. Compared to the simulation with the MYJ scheme, the TC size simulated with the YSU scheme was even larger, leading to a stronger IMF transport from the WPSH to the TC and a weaker WPSH.

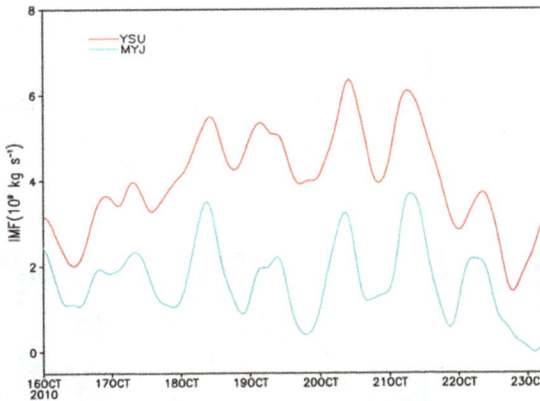

Fig. 5.4. Temporal evolutions of IMF (unit: 10^9 kg·s^{-1}) within a radius of 800 km from the TC center below 1 km simulated with YSU scheme (red line) and MYJ scheme (blue line), respectively (positive inward).

The above discussion indicates clearly that the TC size simulated with the two different PBL schemes is different, which subsequently results in different IMFs from the WPSH to the central TC area. Corresponding to the different IMFs, the WPSH intensity changed in opposite direction in the simulations with the SYU and MYJ schemes, and the large-scale steering flow changed differently in response to the different WPSH changes, eventually leading to large differences in the simulated TC track. In order to further explore the mechanisms for the impact of PBL scheme on TC simulation, in the next section we analyze the difference in the description of PBL processes between the two schemes, and reveal the reasons for the difference in the TC size simulation using the two schemes.

5.4 Spatial pattern of water vapor distribution

The PBL scheme affects the model simulation via describing the vertical transport of physical variables (e.g., water vapor, heat flux, etc.) and latent heat release in the PBL [Wang *et al.*, 2014]. Various PBL schemes based on different theories could result in significant differences in the simulated PBL structure [Huang *et al.*,

2014]. Fig. 5.5 presents the time-height cross sections of rain water and cloud water mixing ratios within the radius of 500km from the TC center simulated with the YSU and MYJ schemes before the TC turned abnormally early. There existed no large differences between the two simulations before 0000 UTC 18 October. High rain water and cloud water contents occurred in the upper troposphere with the large value center located between 500 hPa-450hPa (Fig. 5.5a and 5.5b); no large differences in cloud water and ice water were found between the two simulations except that the area of high water content simulated with the YSU scheme was slightly larger than that with the MYJ scheme, and could water content was smaller than rain water content in both simulations (Fig. 5.5c and 5.5d). After 0000 UTC 18 October, large differences in the water content occurred between the two simulations.

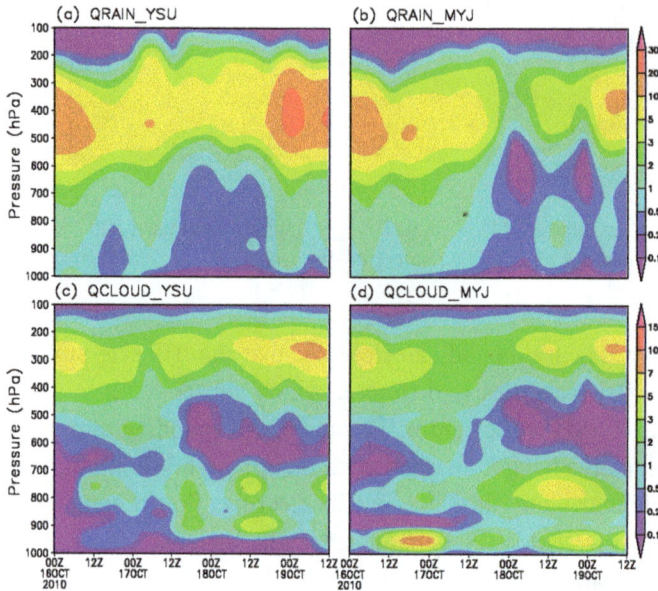

Fig. 5.5. Height-time cross sections of the rain water (a, b) and cloud water mixing ratios (c, d) (unit: 10^{-2} g·kg^{-1}) averaged within a radius of 500 km from the TC center during the period of the TC track departure simulated with YSU scheme (a, c) and MYJ scheme (b, d), respectively.

In the upper level, the water content simulated with the YSU scheme was much higher than that simulated with the MYJ scheme. 12 hours later at 1200 UTC 18 October, the simulated TC track became significantly different between the two simulations, and the TC turned northward earlier than observation in the simulation with the YSU scheme. Hong *et al.* [2006] found that the subgrid-scale turbulent mixing caused by the surface heat fluxes is considered in the non-local YSU scheme, which intensifies the turbulent mixing throughout the entire PBL and transports heat and moisture to the upper troposphere. In contrast, turbulent mixing only occurs between two adjacent levels in the local MYJ scheme [Brown, 1996]. Vertical mixing in the PBL is not sufficient enough to transport large amounts of heat and moisture to the upper troposphere [Bright and Mullen, 2002] in the MYJ. The different mixing processes in the two PBL schemes lead to large differences in the water content above the TC region between the two simulations. Compared to the MYJ scheme, the YSU scheme could transport more heat and moisture to the upper levels over the TC region. The latent heat release resulted in abnormal increases in TC intensification and size, and eventually led to the departure of the simulated TC track.

In order to explain differences in the vertical moisture transport between the two schemes, Fig. 5.6 presents the vertical water vapor flux transport at 850 hPa (positive upwards) simulated with the YSU and MYJ schemes at three times before and after the abnormal turning of TC in the simulation with YSU. The results indicate that during the TC development period, pronounced upward water vapor transport occurred in the lower troposphere. In the central TC area, the vertical water vapor transport simulated with the two schemes was similar, although the simulated TC structure was different. In the peripheral region, however, large differences in vertical water vapor transport occurred between simulations with the two schemes. At 0000 UTC 17 October, an obvious water vapor transport path existed to the southeast of the TC in the simulation with the YSU, covering a broad area with large amounts of water vapor transport. In the simulation with the MYJ scheme, a water vapor transport path also existed in the peripheral region of the TC, but it only covered a small area with limited transport of water vapor. Since 1200 UTC 18 October when the simulated TC with the YSU scheme turned earlier than observation, the range of vertical water vapor transport and difference between the simulations with the two schemes gradually increased with several belts of vertical water vapor transport. The area and intensity of vertical water vapor transport to the northeast and southwest of the TC in the simulation with the YSU scheme both were larger than that with the MYJ scheme. By 1200 UTC 20 October, two water vapor transport paths were located to the northeast and southwest of the TC in the simulation with the YSU scheme, and both paths were stronger than that in the simulation with the MYJ scheme, which showed a limited vertical water vapor transport over a small area around the TC center. In summary, the vertical water vapor transport was much stronger in the simulation with the YSU scheme compared to that with the MYJ scheme, particularly in the peripheral region.

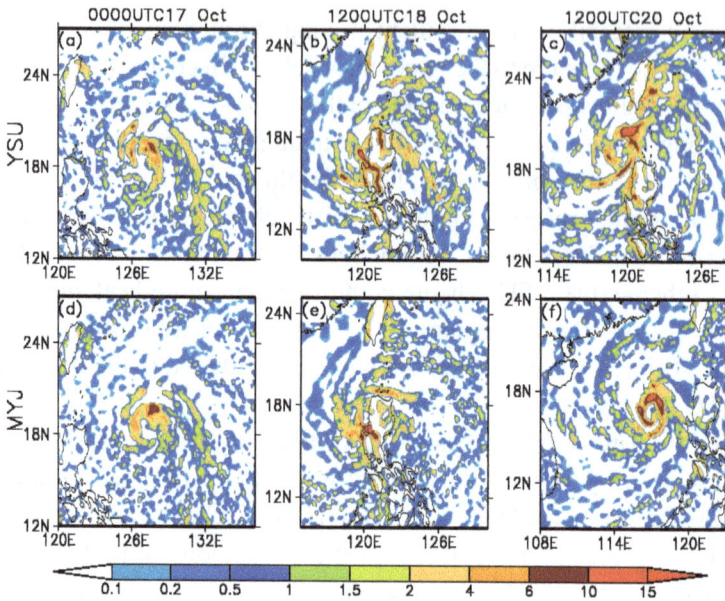

Fig. 5.6. Vertical transport of moisture flux (unit: $g \cdot kg^{-1} \cdot m \cdot s^{-1}$) at 850 hPa at 0000UTC 17 October (a, d), 1200UTC 18 October (b, e) and 1200UTC 20 October (c, f) simulated with YSU scheme (a, b, c) and MYJ scheme (d, e, f), respectively.

Min *et al.* [2010] pointed out that the water vapor supply in the spiral rainbelt region of TC is closely associated with the water vapor convergence in the PBL below 850 hPa. Thereby vertical water vapor transport in the PBL could promote convective development in the spiral rainbelt region and enhance convective activity. Compared to the MYJ scheme, the YSU scheme yielded much stronger vertical water vapor transport in the peripheral region of TC, which was favorable for convective development and more convective activities in the rainbelt region.

Guinn and Schubert [1993] indicated that the TC size might be correlated with activities in the spiral rainbelt region. Kimball [2006] also found that enhanced environmental humility could promote the formation of rainbelt and lead to larger TC size. Fig. 5.7 displays the radar reflectivity simulated with the two schemes and the retrievals of T-3B42 at three times before and after the abnormal turning of TC in the simulation with the YSU scheme. AT 0000 UTC 17 October, the radar reflectivities simulated with the two schemes were already different (Figs. 5.7a and 5.7d). The rainbelt in the peripheral region simulated with the YSU scheme was broad and a complete spiral rainbelt appeared to the southeast of the TC. In the simulation with the MYJ scheme, however, a strong rainbelt appeared to the north of the TC while a few rainbelts occurred to the southeast of the TC, which was consistent with observations shown in the retrieval of T-3B42 (Fig.

5.7g). At 1200 UTC 18 October, the reflectivity simulated by the YSU scheme intensified distinctly, and the rainbelt area in the peripheral region increased significantly. An abnormally strong rainbelt occurred to the northeast of the TC (Fig. 5.7b). In the simulation with the MYJ scheme, the rainbelt remained near the TC center while only a few weak rainfall areas occurred to the northeast of the TC (Fig. 5.7e), which was consistent with observations (Fig. 5.7h). At 1200 UTC 20 October after the early turning of TC in the simulation with the YSU scheme, a large spurious heavy rainfall belt occurred to the north of the TC (Fig. 5.7c). In the simulation with the MYJ scheme, the rainbelt distribution was similar to observations with large precipitation concentrated over the central TC region while the TC eye without reflectivity was also obvious (Fig. 5.7f and 5.7i).

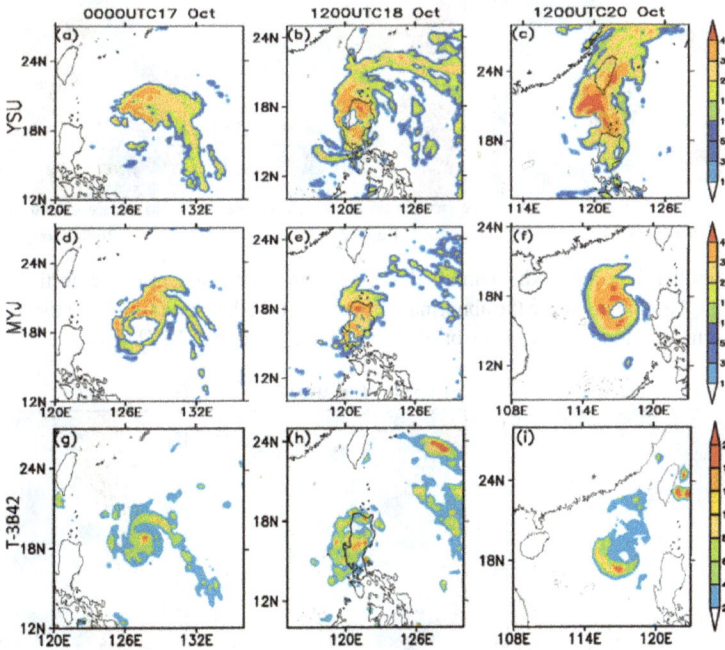

Fig. 5.7. Simulated radar reflectivity (unit: dBZ) at (a, d) 0000UTC 17 Oct., (b, e) 1200UTC 18 Oct.18 (c, f) 1200UTC 20 Oct. and T-3B42 retrieved precipitation rate (unit: mm·hr^{-1}). (a, b, c) : YSU scheme; (d, e, f) : MYJ scheme; (g, h, i) : T-3B42 retrieved precipitation rate.

Figs. 5.8 and 5.9 show the azimuthal-averaged pressure gradient (positive inward) and winds to the north of TC center (positive outward) at 500 m in the simulations with the two PBL schemes and their differences. The radial pressure gradient simulated with the YSU scheme was smaller than that simulated with MYJ within the radius of 350km from the TC center except the inner core region of the TC (Fig. 5.8c). However, outside the radius of 350km from the TC center, the radial pressure gradient in the YSU simulation was larger than that in the MYJ simulation, suggesting that the IMFs from the WPSH to TC were larger in the YSU simulation than in the MYJ simulation over the area to the north of the TC (Fig. 5.9c). Similarly, the radial velocity in the YSU simulation was smaller than that in the MYJ simulation within the radius of 350km from the TC center except the inner core region, whereas it was larger outside the radius of 350km (Fig. 5.9c).

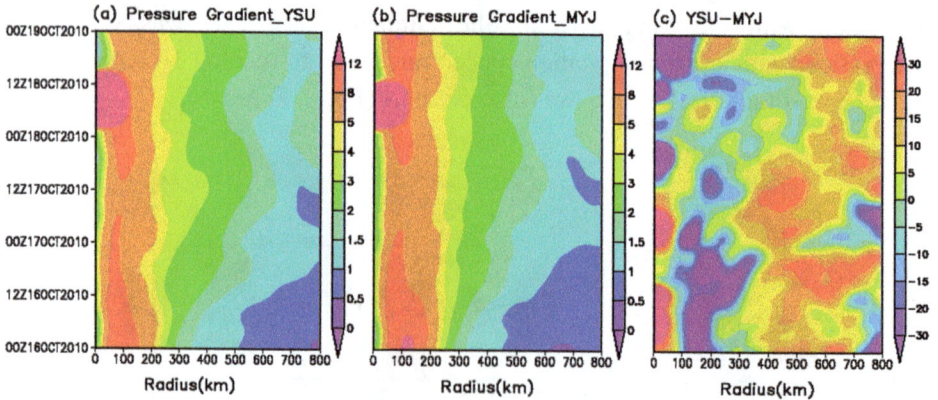

Fig. 5.8. Hovmöller diagram of the azimuthal-averaged pressure gradient to the north of TC center (unit: 10^{-3} Pa·m^{-1}) at 500 m before the abnormal turning of TC in the simulation with the YSU scheme (a), MYJ scheme (b), and percentage error of YSU scheme relative to MYJ scheme (c, %).

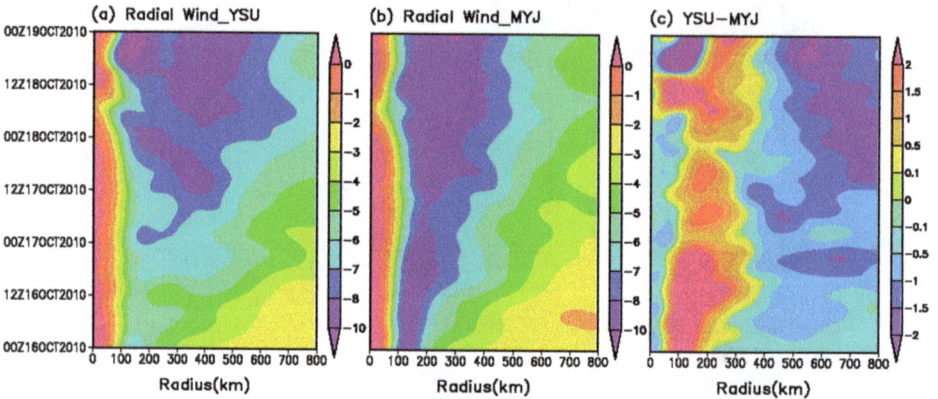

Fig. 5.9. Hovmöller diagram of the azimuthal-averaged radial wind to the north of TC center (unit: m·s^{-1}) at 500 m before the abnormal turning of TC in the simulation with YSU scheme (a), MYJ scheme (b), and their differences (c).

As a result, the spiral rainbelt in the peripheral region was more active in the YUS simulation. The studies of Hill and Lakmann [2009] and Xu and Wang [2010a, 2010b] both suggested that an active spiral rainbelt could lead to more release of latent heat flux and expand the periphery of TC to a larger area, leading to increases in TC size. In the YSU simulation, the rainbelt in the peripheral region was more active with larger latent heat release than that in the MYJ simulation, leading to significant decreases in local pressure and weakened the inward pressure gradient and radial velocity. Eventually the simulated TC size was larger in the YSU simulation compared to that in the MYJ simulation. A larger TC size increased the IMF transport from the southern flank of the WPSH to the TC center, which weakened the TC intensity and resulted in early turning of the TC.

5.5 Conclusions

It is found that TC track simulations with both two PBL schemes could well capture the westward moving features of Megi at its early stage. However, after Megi made landfall in the northern Philippines, the experiment with the MYJ scheme could still well reproduce the continuous westward moving of Megi and its abrupt turning to the north and subsequent landing in Zhangpu County in northern Fujian. Thereby, it could be inferred that the interaction between Megi and the large-scale circulation was realistically simulated with the MYJ scheme. However, in the simulation with the YSU scheme, the spiral rainbelt of Megi developed abnormally around the time when Megi made landfall in the northern Philippines; the TC size increased correspondingly, and the simulated TC turned northward earlier than observation.

In addition, the vertical mixing within the PBL was stronger in the simulation with the YSU than that with the MYJ scheme. Under appropriate condition, more water vapor was transported to the upper troposphere in the simulation with the YSU scheme, forming hydrometers surrounding the TC. Strong vertical transport of water vapor also intensified convective activities in the rainband region in the periphery of TC, which led to more latent heat release and reduced local pressure, resulting in larger TC size.

As a deep low-pressure system, when it moves along the southern flank of the subtropical high, there exist IMFs from the subtropical high to TC. In the simulation with the YSU scheme, the TC size was larger than that simulated with the MYJ scheme. Correspondingly the pressure gradient also increased in the periphery of the TC, which intensified the IMFs from the subtropical high to TC region. As a result, the subtropical high weakened, leading to changes in the large-scale steering flow and early turning of the TC.

Therefore, the PBL scheme could affect the TC size via modifying the vertical transport of water vapor in the area surrounding the TC. In the present study, the differences in TC size simulated with different PBL schemes were analyzed. It was found that the TC size simulated with the YSU scheme was larger compared to that simulated with the MYJ scheme. Thereby, more IMFs were transported from the subtropical to the TC area, which resulted in changes in the large-scale steering flow and subsequent turning ahead of schedule for Megi.

Chapter 6

Sensitivity of Tropical Cyclone Track to Storm Size

6.1 Introduction

TC impacts are highly correlated with the storm size, yet the importance of storm size has not received enough attention. The TC size is an important structure parameter not only because it determines the extent of the damage caused by the TC [Powell and Reinhold, 2007; Maclay *et al.*, 2008], but also because it has great impacts on the motion of the TC [Lester and Elsberry, 1997; Lester and Elsberry, 2000; Hill and Lackmann, 2009]. Theoretically, the storm size could affect storm motion by influencing the extension and intensity of anvil clouds [Bu *et al.*, 2014] or by influencing the outer wind structure. Yet which influence plays a major role in the storm motion remains unknown. As discussed in Sun *et al.* [2014a] and Sun *et al.* [2015b], the anvil clouds could change the microphysical latent heating over the edge of the WPSH, which subsequently affects the WPSH and TC motion. Meanwhile, the TC movement often deviates from the large-scale steering flow due to the beta-effect propagation (BEP), which depends on the mean relative angular momentum and thus is highly sensitive to the outer wind structure of a TC [Holland, 1983; Fiorino and Elsberry, 1989; Carr and Elsberry, 1997]. For this reason, the movement of large storms may differ from that of smaller ones due to the more pronounced beta drift [Hill and Lackmann, 2009]. Observational analyses have also confirmed the relationship between the TC track and TC size. Lee *et al.* [2010] calculated the size of 145 TCs in the WNP during 2000-2005 based on the QuikSCAT oceanic winds and the best tracks of the TCs from the JTWC. Their results indicate that the 18 persistently large TCs mostly have northwestward or north-northwestward tracks, while the 16 persistently small TCs mostly move westward to northwestward. However, due to the lack of observations with a wide variety and a high spatial-temporal resolution, it is hard to reveal the mechanism behind the observed phenomena.

While most previous studies on TC size have focused on TC internal processes and interactions between the TC and environmental circulations and their impact on TC size [Lee *et al.*, 2010], the present study explores the impact of

TC size on the interaction between the TC track and WPSH instead of changes in the TC size itself. Emanuel [1986] and Rotunno and Emanuel [1987] proposed that the size of the initial disturbance is a key factor in determining the TC size. Following their studies, here we assume that the storm size is closely related to its initial size. We will further explore the impact of TC size on the simulations of TC motion and the WPSH by changing the initial size of the storm.

We will first investigate the impact of the initial storm size on TC motion and the WPSH intensity, and then explore the involved physical processes and possible mechanisms. The selected TC cases are still the Songda (2004) and Megi (2010) as for the studies on the effects of model physics schemes.

6.2 Model configuration and experimental design

To illustrate the impact of initial TC size on TC track and WPSH simulations, we have performed two case studies on Songda (2004) and Megi (2010). Both Songda and Megi are characterized by stronger intensity, long duration, and fast development with a typical turning track. Their motions and sudden turnings are closely related to the withdrawal and extension of the WPSH. The track information of Songda and Megi are provided by Regional Specialized Meteorological Center (RSMC).

Songda is among the strongest typhoons that made landfall on the main islands of Japan in the past 50 years. It caused extensive damages to Japan due to its strong winds. The storm formed in Marshall Islands on 28 August 2004 and rapidly intensified while moving northwestward over the WNP. Because of the weakening subtropical high, Songda turned to the northeast direction over the East China Sea on 1200 UTC 6 September and made landfall on Kyushu Island, south of the main island of Japan, at 0000 UTC 7 September (see Fig. 6.3).

Megi is one of the most intense TCs on record, and is the only super-typhoon in 2010. Megi formed over the WNP (11.9°N, 141.4°E) at 0000 UTC, 13 October 2010. Due to the influence of the subtropical ridge and the favorable environmental condition, Megi started moving westward after its formation and continued to gain strength. It has reached its peak intensity while making landfall over Isabela Province, Philippines at 0325 UTC 18 October. Megi became weak when passing Sierra Madre due to the effects of the land surface but rapidly regained strength over the SCS. Later on 19 October, Megi turned northwestward and moved slowly since the subtropical ridge weakened due to a deep mid-latitude shortwave trough that was approaching. On October 23, Megi weakened to a tropical storm as it made landfall at Zhangpu in Fujian Province, China. Megi further downgraded to a tropical depression later on October 23 (see Fig. 6.4).

The model used in this study is the Advanced Research version of Weather Research and Forecasting Model, version 3.3.1 (WRF-ARW V3.3.1) developed at the National Center for Atmospheric Research [Skamarock *et al.*, 2008]. WRF-ARW is a three-dimensional, fully compressible, nonhydrostatic model formulated in a terrain-following mass coordinate in the vertical. The National Center for Environmental Prediction (NCEP) global final (FNL) analysis data at $1°×1°$ latitude-longitude grids with 6-h interval is used to provide initial and lateral boundary conditions for the WRF-ARW model. The model configuration for the simulation of TC Megi (2010) is identical to that in our previously study [Sun *et al.*, 2015b] except that the initial time is different, as follows in the next two paragraphs. A 20-km resolution domain with 36 vertical levels is set up for the simulations of both Songda and Megi. Note that the model domains and simulation time for the two cases are different. For the case of Songda, the model domain is centered at (28°N, 137.5°E) with 206 (north-south) × 222 (east-west) grid points and the simulation is initialized at 0000 UTC 31 August and ends at 0600 UTC 07 September 2004, covering a total of 174 hours. For the case of Megi, the model domain is centered at (22°N, 122°E) with 160 (north-south) × 180 (east-west) horizontal grid points and the simulation is initialized at 0000 UTC 16 October and ends at 0000 UTC 24 October 2010, with a total of 192-hour integration. The domains of the two cases all extend far enough south to allow simulations of the WPSH withdrawal and the recurvature of the TCs.

The model physics used in this study include (i) the single-moment 3-class microphysics scheme [Hong *et al.*, 2004]; (ii) the Grell-Dévényi (GD) cumulus parameterization scheme [Grell and Dévényi, 2002]; (iii) the Mellor-Yamada-Janjić boundary layer scheme [Mellor and Yamada, 1982; Janjić, 2002] with the Monin-Obukhov surface layer scheme [Janjić, 1996, 2002]; (iv) the 5-layer thermal diffusion scheme for land surface processes [Skamarock *et al.*, 2008]; and (v) the Goddard scheme for shortwave radiation calculation [Chou and Suarez, 1994] and Rapid Radiative Transfer Model (RRTM) for longwave radiation calculation [Mlawer *et al.*, 1997]. For each TC case, three experiments with different initial storm sizes are conducted to investigate the response of TC track and the WPSH to changes in initial TC size. Here, the TC Bogus scheme in the WRF model is used to change the maximum radius from TC center at the initial time [Skamarock *et al.*, 2008]. In these experiments, the maximum radius outward from the TC center is set to 60-, 120-, and 180-km at the initial time, respectively. For convenience, we define the three experiments as the one with a small-sized storm (ES), the one with a medium-sized storm (EM), and the experiment with a large-sized storm (EL) in order of increasing size, respectively. All other physical schemes and model settings are the same in the three experiments described above.

6.3 Simulation results

6.3.1 *Storm size*

In an operational setting, storm size is described by the area of the outermost closed isobar (ACI) in the surface level. For both Songda and Megi, the value of the outermost closed isobar is about 1000 hPa. Fig. 6.1 shows the temporal evolutions of ACI in the cases of Songda and Megi. It clearly shows that, in both Songda and Megi cases, the ACI is highly sensitive to the initial vortex size determined by the TC Bogus scheme, and increases significantly as the initial vortex size increases, especially as the initial vortex size increases from small size in the ES to medium size in the EM. This is consistent with the idealized model results of Xu and Wang [2010a], which indicated that a storm with a large initial size usually has strong outer winds and large surface entropy fluxes outside the eyewall. They are accompanied by active spiral rainbands, leading to fast increase in the inner-core size. In addition, the ACI in the EL decreases significantly after 2000 UTC 04 September 2004 for Songda case. This is probably caused by the landfall of the storm in the EL, since the time of the decrease in ACI is basically consistent with the landfall time of the storm simulated in the EL (see Fig. 6.3).

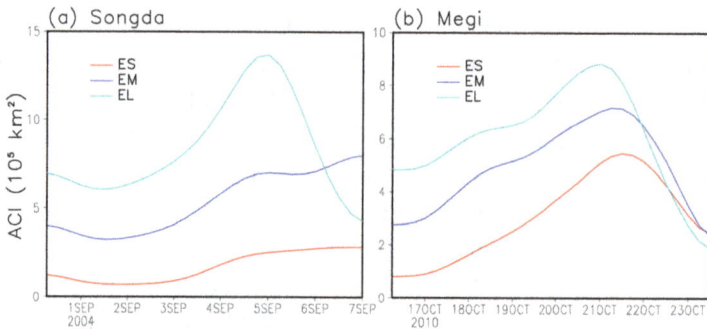

Fig. 6.1. Temporal evolutions of ACI in the sensitivity experiments in (a) Songda case and (b) Megi case.

To further provide a picture of typical precipitation associated with the simulated TC, Fig. 6.2 presents two snapshots of the model-simulated radar reflectivity at 0000 UTC 3 September 2004 for the case of Songda and 0000 UTC 18 October 2010 for the case of Megi. Compared with that in the ES run, a wider and broader eyewall is evident in the EM and EL runs, along with larger area of precipitation in the outer spiral rainbands in both Songda and Megi cases. This is consistent with our hypothesis and further indicates that the size of the simulated

TCs is highly dependent on the initial vortex size, i.e. the larger the initial vortexes, the larger the storms will be later.

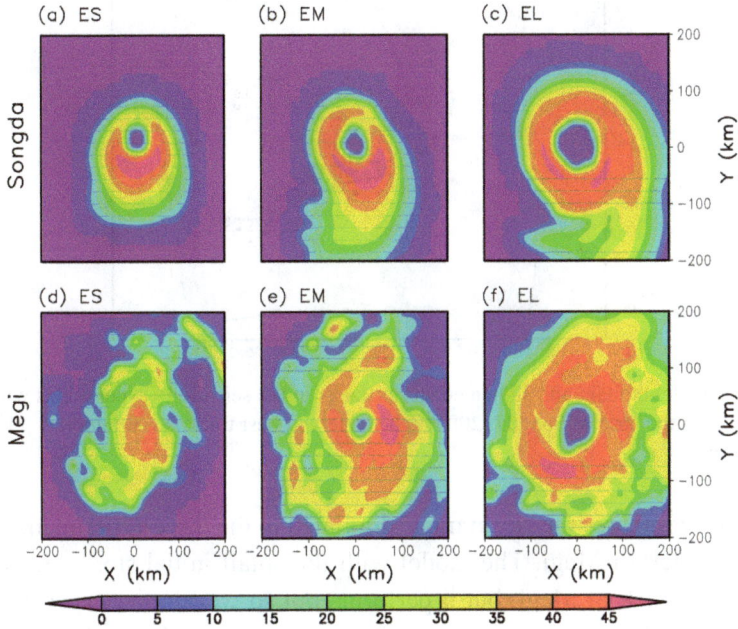

Fig. 6.2. Module-simulated radar reflectivity (unit: dBZ) for the sensitivity experiments at 0000 UTC 3 September 2004 in the case study of Songda (2004) (a, b, c) and 0000 UTC 18 October 2010 in the case study of Megi (2010) (d, e, f), respectively.

6.3.2 *Storm track*

Previous studies have indicated that the storm size affects its motion not only by changing the large-scale environmental flow, but also by influencing the BEP [Holland, 1983; Fiorino and Elsberry, 1989; Carr and Elsberry, 1997; Bu *et al.*, 2014; Sun *et al.*, 2014a; Sun *et al.*, 2015c]. Fig. 6.3 compares the storm track simulated in the sensitivity experiments with the JTWC best track of Songda. It indicates that the simulated storm track is highly sensitive to the initial size of the storm and large differences between the results of the three experiments occur about two days after the model integration starts. The simulated storm in the EM and EL turns northward earlier than observation and makes landfall in Japan at about 1200 UTC 06 September 2004 and 1800 UTC 05 September 2004 respectively, whereas the simulated storm in the ES continues to move westward and turns northward later than observation and didn't make landfall before 0600 UTC 07 September 2004.

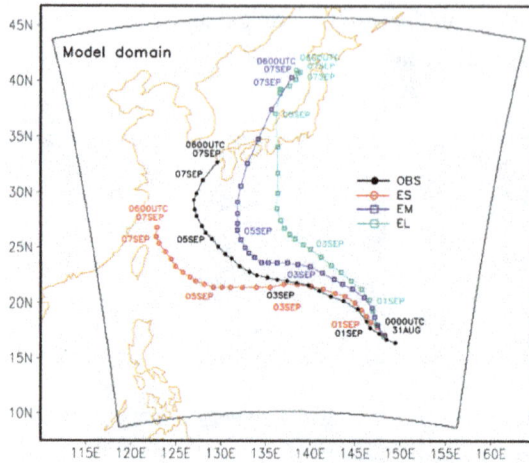

Fig. 6.3. The model domain and simulated storm tracks in the sensitivity experiments with varying initial TC size in the case of Songda (2004). The observed best track at 6-h intervals (black dotted line) is overlaid.

Fig. 6.4 compares the storm track simulated in the three experiments with the JTWC best track of Megi. The model with the small initial storm size can well reproduce the track of Megi, but it performs not so well in the experiments with the medium and large initial size. All experiments realistically simulate the northwestward movement of Megi before 0000 UTC 17 October 2010 and the west-southwestward movement along the southern periphery of the WPSH until the storm crossed the Luzon Island. Large differences between results of the three experiments occur after 1800 UTC 18 October 2010. Similar to the results in Songda case, the simulated storm in the EM and EL turns northward earlier than observation, whereas in ES it continues to move westward and turns northward over SCS at about 1800 UTC 19 October 2010. Apparently, the simulated storm track is sensitive to the initial size of the storm. It is worth noting that a TC with a larger initial size turns northward earlier in both Megi and Songda cases.

Feedback and interaction between the TC and WPSH are interwoven in these simulations, making it a challenging issue to address what is the root cause of the large biases in both the WPSH and TC simulations. Note that the unrealistic withdrawal and extension of the simulated WPSH is responsible for the failure in RCM simulations of TC motions. The erratic departure of the simulated TC track from its observed position contributes greatly to the RCM's failure in simulating the WPSH [Zhong, 2006]. Thereby, the TC track simulation is a key factor that affects the WPSH simulation. In the following section, we will discuss in detail the possible reasons for the difference in TC track simulation between these experiments.

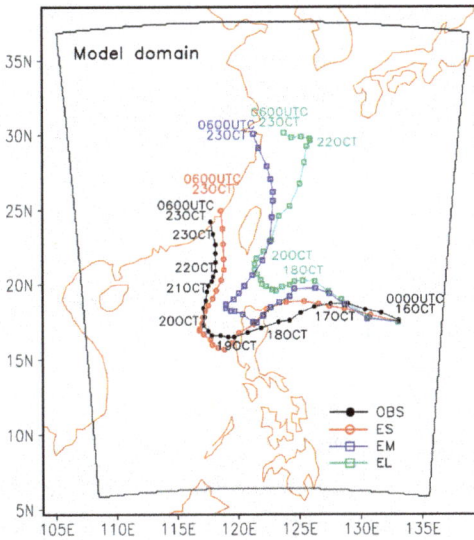

Fig. 6.4. As in Fig. 6.3, but for the case of Megi (2010).

6.4 Possible reasons for the differences in TC tracks

6.4.1 *Potential vorticity tendency diagnosis*

Previous studies suggested that the environmental flow and the TC structure are two key factors determining the TC motion over the WNP [e.g., Chan and Gray, 1982; Holland, 1983; Fiorino and Elsberry, 1989; Wu and Wang, 2000; Wu *et al.*, 2005; Zhong, 2006]. Theoretically, the storm size can influence the TC motion in two ways. First, the storm size can modulate the large-scale environmental flow near the TC and thus affect TC motion by influencing the withdrawal and extension of the WPSH. Second, the storm size can affect the TC motion by modifying the thermodynamic and dynamic structure of the TC. As suggested by Carr and Elsberry [1990] and Holland [1993], the large-scale environmental flow is defined as the layer-mean (850-300 hPa) flow averaged over a 5°-7° latitude band of the TC center. In the following paragraph, we will discuss which factor is dominant in the differences between the simulated TC motions of the three sensitivity experiments.

To estimate contributions of the TC structure and environmental flow to TC motion, the potential vorticity tendency (PVT) diagnosis technique is applied as in Chapter 4, namely that proposed by Chan [1984] and Wu and Wang [2000]. Simulations of an ideal TC indicated that a baroclinic TC moves towards the region where the azimuthal wavenumber 1 of the maximum PVT is located, and it is suggested that the PVT results from horizontal PV advection (HA), vertical PV transportation (VT), and diabatic heating (DH), while the contribution of individual physical process to the TC motion is equivalent to its contribution to the wavenumber 1 component of the PVT [Wu and Wang. 2000]. The corresponding diagnosis equations are same as Eq. (4.1) – Eq. (4.4) and the representation of symbols could see in Chapter 4 in detail.

For the convenience to interpret the results, we compare PVT between the EM and EL simulations. Note that for Megi case, the assumption of Eq. (6.1) is no longer valid in ES due to the reduced intensity of the symmetric TC circulation and the enhanced asymmetric circulations. A large bias in the estimated speed (i.e., V_{PV}) occurs in the ES before 1200 UTC 18 October 2010. Thereby, we compare the results of PVT between the EM and EL in both Megi and Songda cases.

Fig. 6.5 depicts the temporal evolution of the vertically averaged TC moving speed calculated from the PVT equations as well as the individual contributions of HA, VT and DH calculated in the EM and EL for the case of Songda (2004). The calculation covers the time period from Sep. 1 to Sep. 5 2004, corresponding to the period before and after the time when the difference in TC position between the EM and EL becomes significant. The mean moving speed of the TC is averaged between 850 hPa and 400 hPa because of the various vertical extents of the positive PV anomalies in the EM and EL. As is shown, the mean moving speed of TC calculated from the wavenumber 1 component of the PVT (V_{PV}) is consistent with that calculated from the TC center position (V_C). A similar result can be found in previous studies [Wu and Wang, 2000; Sun *et al.*, 2015b]. Thereby, the PVT diagnosis approach has been proved to be an effective method to estimate the TC moving speed for real TC cases.

Based on Eq. (4.2), the contribution of each individual physical process (i.e., HA, VT, and DH) to the difference in TC motion between the EM and EL can be estimated. In the following analysis, we will focus on the period before 1200 UTC 03 September, when the difference in TC position between the EL and EM is not significant for Songda case. This period is selected to ensure the background circulation is the same or similar for the EM and EL experiments. Once the difference in TC position becomes evident, the effects of individual physical process in Eq. (4.2) will be susceptible to differences in the background environment and thus cannot be used for comparison.

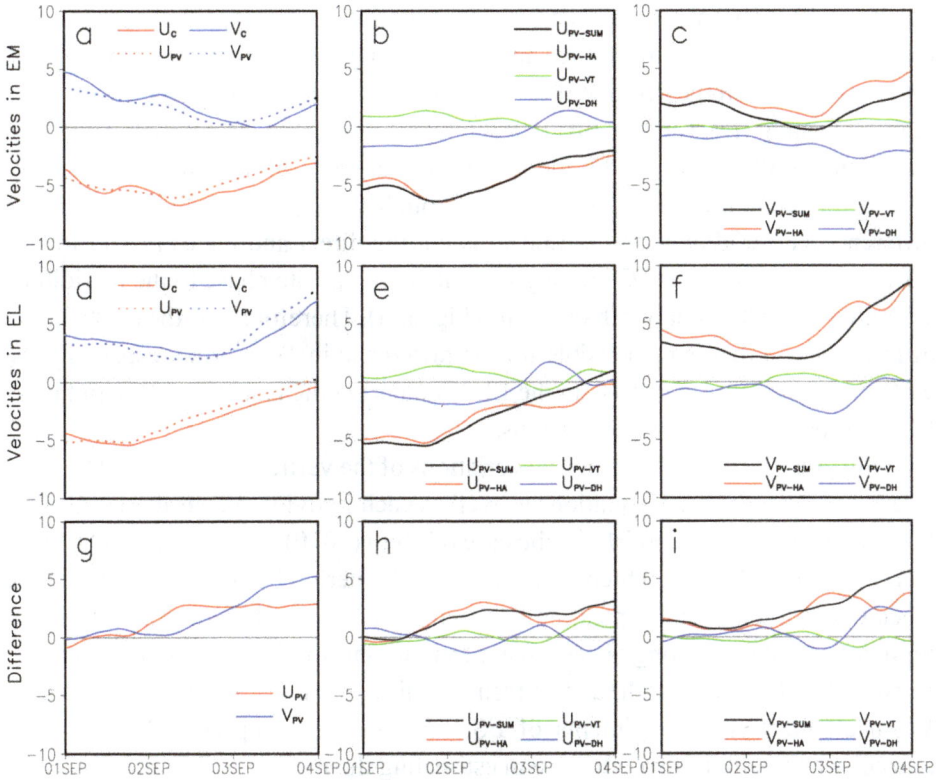

Fig. 6.5. Vertical mean zonal and meridional TC motion speed (m s^{-1}) calculated from the center position ($\mathbf{V_C}$), the PV tendency ($\mathbf{V_{PV}}$), and individual contributions of HA, VA, DH ($\mathbf{V_{PV\text{-}HA}}$, $\mathbf{V_{PV\text{-}VT}}$ and $\mathbf{V_{PV\text{-}DH}}$) and their summation ($\mathbf{V_{PV\text{-}SUM}} = \mathbf{V_{PV\text{-}HA}} + \mathbf{V_{PV\text{-}VT}} + \mathbf{V_{PV\text{-}DH}}$) in EM (a-c; top panel), EL (d-f; middle panel), and their difference (EL-EM) (g-i; bottom panel) in the case of Songda (2004). All calculations are averaged within a radius of 360 km from TC center and between the levels 850 hPa and 400 hPa.

Due to the strong large-scale forcing of the WPSH in the EM, the contribution of HA ($\mathbf{V_{PV\text{-}HA}}$) is notably stronger than that in EL, resulting in a large difference in the zonal TC moving speed especially after 0000 UTC 02 September 2004 (Figs. 6.5b and 6.5e). Similar to the zonal wind impact on TC motion, the meridional wind also contributes greatly to the TC motion along the meridional direction (Figs. 6.5c and 6.5f). More importantly, the magnitude of zonal $\mathbf{V_{PV\text{-}HA}}$ in the EL is notably smaller than that in EM by up to -3 m s^{-1}, but the magnitude of meridional $\mathbf{V_{PV\text{-}HA}}$ in the EL is much larger than that in the EM by up to 4 m s^{-1}. This suggests that $\mathbf{V_{PV\text{-}HA}}$ plays an important role in reducing the westward moving speed of TC and accelerating its northward moving speed in the EL. This is consistent with the results of Bi *et al.* [2015], which also emphasized the important impact of horizontal vorticity advection on TC motion. Similar to the results in

Sun *et al.* [2015b], $\mathbf{V}_{PV\text{-}VT}$ makes little contribution to the TC moving speed and thus causes almost no difference in TC motion between the two experiments. Previous studies suggested that there is a fast adjustment between the asymmetric diabatic heating (DH) and relatively asymmetric flow (HA) [Peng *et al.*, 1999; Wu and Wang, 2000]. This is why the temporal evolutions of $\mathbf{V}_{PV\text{-}DH}$ and $\mathbf{V}_{PV\text{-}HA}$ are anti-correlated in meridional direction. Although $\mathbf{V}_{PV\text{-}DH}$ contributes greatly to the difference in meridional TC motion, it is not the direct and major reason for the difference in meridional TC moving since it is of opposite phase to the meridional TC moving and it is smaller than $\mathbf{V}_{PV\text{-}HA}$ (Fig. 6.5i). Therefore, it is the contribution of HA ($\mathbf{V}_{PV\text{-}HA}$) that is responsible for the difference in TC motion, especially the early northward-turning of TC in the EL. Comparing the results in ES with that in EM, we can reach similar conclusions.

Fig. 6.6 illustrates the temporal variations of the vertically averaged TC speed calculated from the PVT equations as well as each individual contribution of HA, VT and DH in the EM and EL for the case of Megi (2010). The time period is from Oct. 17 to Oct. 21 2010, which corresponds to the period before and after the time when the difference in TC position between the EM and EL becomes significant. Similar to studies for Songda case, the mean moving speed of TC is also averaged between 850 hPa and 400 hPa. The results indicate that $\mathbf{V}_{PV}$ is almost identical to $\mathbf{V}_C$ in all the three experiments of ES, EM, and EL, implying that the PVT diagnosis approach is also effective in estimating the motion of TC for Megi case.

In the following analysis for the case of Megi, we focus on the period before 0000 UTC 19 October 2010 when the difference in TC position between the EM and EL is not significant. Similar to the results in Songda case, differences in the contribution of HA ($\mathbf{V}_{PV\text{-}HA}$) to the TC motion is primarily responsible for the difference in the simulated TC motion between the EM and EL (Figs. 6.6h and 6.6i). $\mathbf{V}_{PV\text{-}HA}$ contributes greatly to early northward-turning of the TC in the EL experiment (Fig. 6. 6i).

Following the approach in Wu and Wang [2000] (a case study of ideal TC), we have considered the vortex as a symmetric PV anomaly in Eq. (4.1). However, this assumption may not hold strictly because significant asymmetric PV anomalies might exist in real TC cases, which contributes to the notable difference between $\mathbf{V}_{PV\text{-}SUM}$ in Fig. 6.6f (Fig. 6.6c) and $\mathbf{V}_{PV}$ in Fig. 6.6d (Fig. 6.6a). Nevertheless, this difference does not affect the reliability of our main conclusions. Due to the significant asymmetry of TC in the Megi case (Figs. 6.2d-6.2f), $\mathbf{V}_{PV\text{-}SUM}$ in Figs. 6.6f (Fig. 6.6c) are underestimated. However, the difference in $\mathbf{V}_{PV\text{-}SUM}$ between the EL and EM (Fig. 6.6i) is basically consistent with the difference in $\mathbf{V}_{PV}$ between the EL and EM (Fig. 6.6g). We make conclusions based on the difference between EL and EM simulations rather than on the individual result of

EM or EL. Together with the difference in $\mathbf{V}_{PV-HA}$ between EL and EM (Fig. 6.6i), all the different results between EL and EM clearly indicate that it is the contribution of HA ($\mathbf{V}_{PV-HA}$) that is responsible for the difference in TC motion between the EL and EM. As suggested by Wu and Wang [2000], $\mathbf{V}_{PV-HA}$ not only includes the contribution of BEP but also includes the contribution of the environmental flow. Thus, the storm size affects TC motion by changing the BEP and the environmental flow near the TC. Next, we will discuss which one is primarily responsible for the large difference in TC motion between the sensitivity experiments.

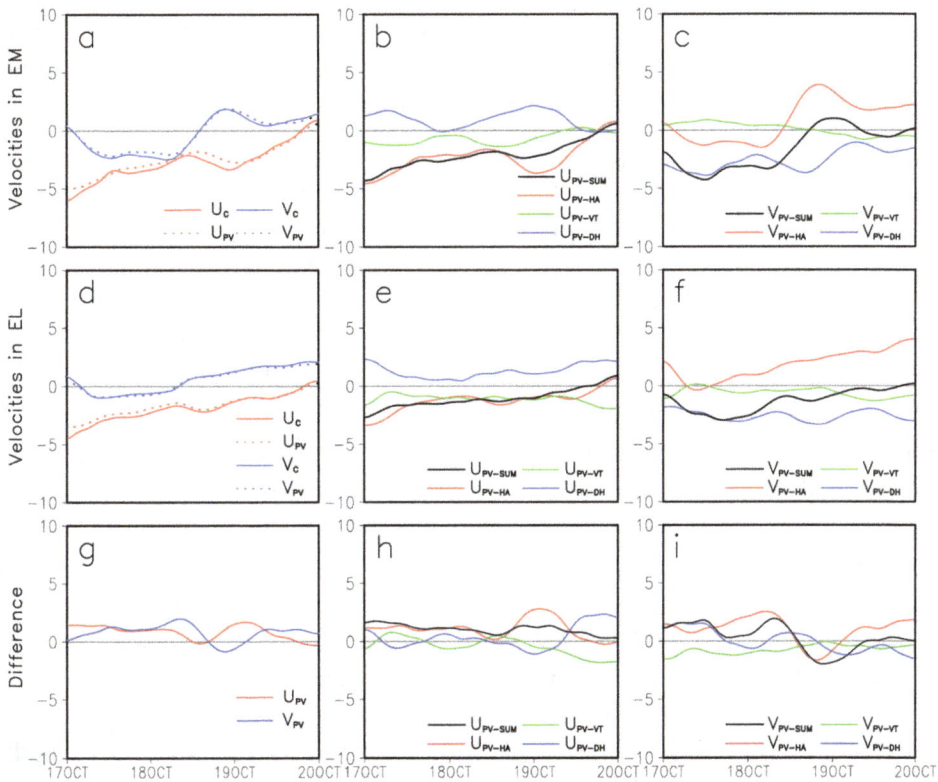

Fig. 6.6. As in Fig. 6.5, but for the case of Megi (2010).

6.4.2 *Contribution of beta-effect propagation on the turning of TC*

To further investigate the contribution of BEP to the meridional TC motion and thus the northward turning of the TC, we calculate the meridional BEP speed based on the relationship between the mean relative angular momentum (MRAM) of the cyclonic circulation and the meridional BEP speed [Wang and Li, 1992]. The MRAM is defined by

$$\text{MRAM} = \frac{\int_{p_1}^{p_0}\int_A v_\lambda(r)rdAdp}{\int_{p_1}^{p_0}\int_A dAdp} \qquad (6.1)$$

where r denotes the radial distance from the vortex center, $v_\lambda(r)$ is the tangential wind of the vortex, A is the horizontal area occupied by the vortex flow at level p, p_1 and p_0 are the upper and lower boundaries of the vortex circulation, respectively.

The relationship between the MRAM (in units of 10^6 m^2 s^{-1}) and the meridional beta-drift speed (m s^{-1}) is essentially nonlinear. As suggested by Wang and Li [1992], the northward drift speed is approximated by

$$V_{\text{BEP}} = a\,|\,\text{MRAM}\,|^{1/2} + b \qquad (6.2)$$

where the coefficients a and b are 1.01 and -0.11 respectively for cyclones. They also suggested that the westward-drift speed exhibits a weak linear correlation with the absolute value of MRAM.

Fig. 6.7 shows the vertically mean meridional TC motion speed calculated from the contribution of HA ($V_{\text{PV-HA}}$) and the meridional BEP speed (V_{BEP}) in the cases of Songda and Megi. While $V_{\text{PV-HA}}$ displays significant changes over time, V_{BEP} tends to be almost stable with an insignificant change and remains at roughly 2 m s^{-1} in all these sensitivity experiments (Figs. 6.7a, 6.7b, 6.7d, and 6.7e). The results of Songda and Megi cases all show that V_{BEP} in the EL is slightly larger than that in the EM due to the larger storm size in EL. However, as the difference in V_{BEP} is much smaller than the difference in $V_{\text{PV-HA}}$ in most of the time, the difference in V_{BEP} contributes little to the difference in $V_{\text{PV-HA}}$ and thus has little impact on TC motion between the EL and EM (Figs. 6.5i and 6.6i). This indicates that it is not the BEP that is responsible for the large difference in TC motion between the EL and EM, especially the early northward-turning of the TC in both Songda and Megi cases. As mentioned previously, $V_{\text{PV-HA}}$ not only includes the contribution of BEP but also includes the contribution of environmental flow. If the contribution of BEP is excluded, then the large-scale environmental flow plays a critical role in determining TC motion in the EM and EL. Fig. 6.8 shows the meridional steering flow, the meridional TC moving speed, and their difference, in the ES, EM, and EL simulations for the cases of Songda and Megi. Following Carr and Elsberry [1990] and Holland [1993], we use the layer-mean (850-300 hPa) flow averaged over a 5-7° latitude band of the TC center to estimate the large-scale environmental steering flow in this study. Comparing the meridional steering flow with the meridional TC moving speed, we find that, in Songda case, the meridional TC moving speed is basically consistent with the meridional steering flow, and increases significantly when the initial storm size increases. However, in Megi case, the time evolution trend of the meridional TC moving speed is similar to that of the meridional steering flow, but the magnitude of the meridional TC moving speed is different from that of the meridional steering flow. More importantly, compared to the effect of the other factors, the large-scale environmental flow plays a much more important role in determining the TC moving speed in both Songda and Megi cases.

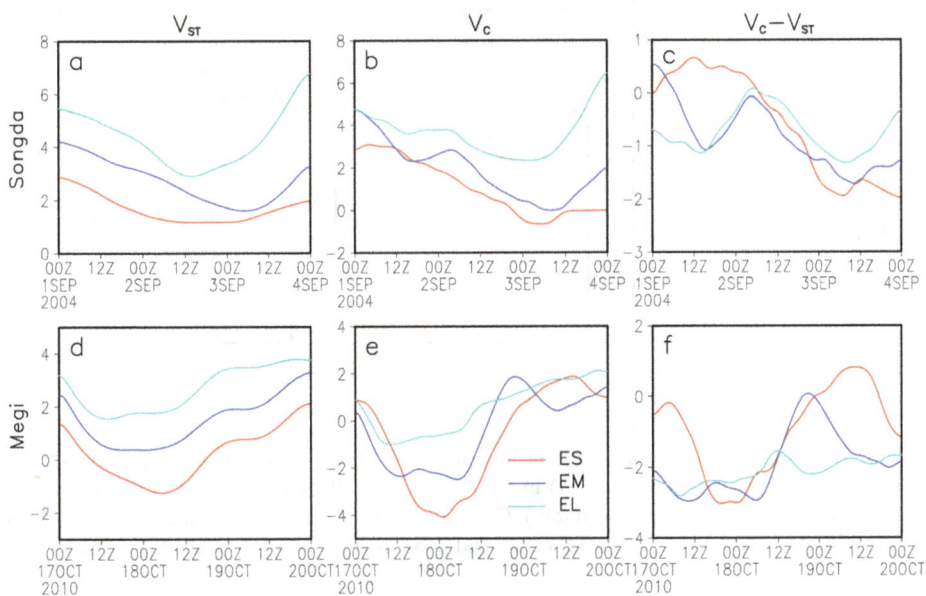

Fig. 6.7. Vertical mean meridional TC motion speed calculated from the contribution of HA (m s^{-1}; $\mathbf{V}_{PV\text{-}HA}$) and the meridional BEP speed (m s^{-1}; $\mathbf{V}_{BEP}$) in EM (left; a, d), EL (middle; b, e), and their difference (right; c, f) in the cases of Songda (top panels) and Megi (bottom panels). All calculations are averaged within a radius of 360 km from TC center and between the levels 850 hPa and 400 hPa.

To further address this issue, the residual term obtained by subtracting the steering flow from the TC moving speed is depicted in Figs. 6.8c and 8f. As is shown, compared with the role of the steering flow, the residual term plays a secondary role in determining the TC motion. The difference in the residual term can be considered as one of the reasons for the difference in TC moving speed between the ES and EM, but it cannot explain the difference in TC moving between the EM and EL. Therefore, it is the large-scale environmental flow that plays a critical role in determining the TC motion. In the following section, we will investigate how the storm size affects the large-scale environmental flow near the TC and finally leads to changes in TC motion.

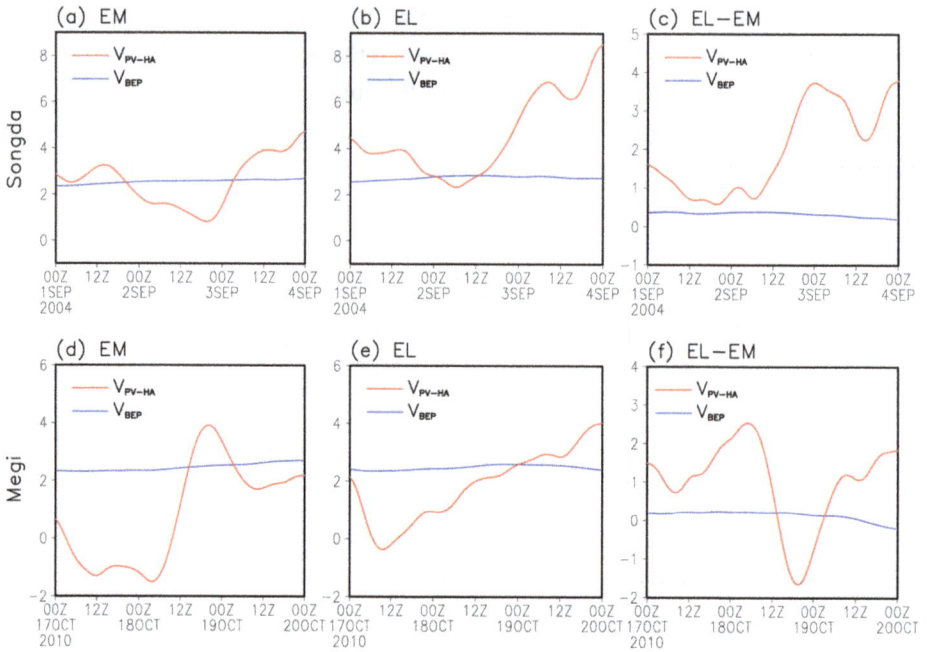

Fig. 6.8. Vertical mean meridional steering flow (m s^{-1}; V_{ST}), meridional TC moving speed (m s^{-1}; V_C) and their difference (m s^{-1}; $V_C - V_{ST}$) in the experiment with different initial storm size in the cases of Songda (upper) and Megi (bottom). Calculations of the steering flow are averaged within the 5-7° latitude radial band from the TC center and between the levels 850 hPa and 300 hPa.

6.4.3 *Effect of large-scale environmental flow on the TC track*

The above analysis has shown that the storm size affects TC motion by changing the environmental flow near the TC (e.g. HA in PVT), which is closely related to the intensity and extent of the WPSH. This indicates that TCs are steered primarily by the large-scale environmental flow, and the characteristics of the TC tracks over the WNP are modulated by the extension and withdrawal of the WPSH. Fig. 6.9 shows the geopotential height at 500 hPa from NCEP reanalysis data and from simulations at 0000 UTC 3 September 2004 and 0000 UTC 18 October 2010, corresponding to the cases of Songda and Megi respectively. In order to clearly represent the difference in the simulated WPSH between the ES, EM, and EL, the geopotential height contour of 5900 m (5880 m) is used to indicate the intensity of

WPSH for Songda (Megi) case in this study. Results of the experiments for both Songda and Megi cases indicate that the strength of the simulated WPSH decreases significantly as the initial storm size increases. Comparing the simulations with the NCEP reanalysis, it is found that larger TCs have more capability to weaken the WPSH and thus are more prone to turn northward, which is consistent with the results of observational analysis in Lee *et al.* [2010] and the conceptual model results in Carr *et al.* [2001]. A comparison of Figs. 6.3, 6.4, and 6.9 suggests that the time and location of the northward turning of the storm is closely related to the degree of the weakening of WPSH due to the strong influence of the steering flow in the southern edge of the WPSH. Actually, the unrealistic early northward turning of the TC simulated in sensitivity experiments (e.g., EM and EL) can be attributed to the unrealistic split of the WPSH. Note that the unrealistic break of the WPSH simulated in these experiments is not caused by the strong storm intensity, since the simulated TC intensity does not increase with the size of the initial storm in these experiments. The simulated TC intensity in EM is even stronger than that in EL for both Songda and Megi cases.

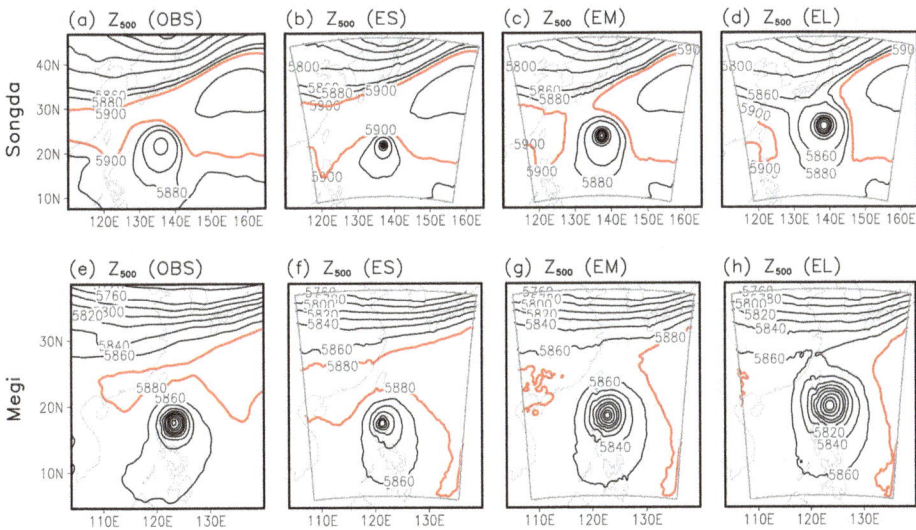

Fig. 6.9. The geopotential height at 500 hPa from NCEP reanalysis data and from simulations at 0000 UTC 3 September 2004 and 0000 UTC 18 October 2010, corresponding to the cases of Songda (2004) (a, b, c) and Megi (2010) (d, e, f) respectively. The contour of 5900 m in Songda case and the contour of 5880 m in Megi case are highlighted in red.

To understand the impact of storm size on the intensity of WPSH, we provide a Hovmöller diagram of the azimuthal-averaged geopotential height at 500 hPa in the TC outer region before the northward turning of the TC in Songda and Megi cases (Fig. 6.10). We only show the radial profiles in TC outer region within a radius of 400-1200 km from TC center. As is shown, there is a semidiurnal cycling in the geopotential height at 500 hPa in both Songda and Megi cases, which may be attributed to the semidiurnal atmospheric tide. More importantly, in both Songda and Megi cases, the simulated geopotential height in the TC outer region decreases significantly as the initial storm size increases, and the contour of the low-value geopotential height (e.g., 5880m in Songda case and 5860m in Megi case) extends outward notably following the increase of the initial storm size. This implies that the simulated intensity of the WPSH on its fringe on the TC side decreases notably with the increase of the initial storm size, which eventually causes the WPSH break in the EM and EL and leads to large differences in TC track simulation.

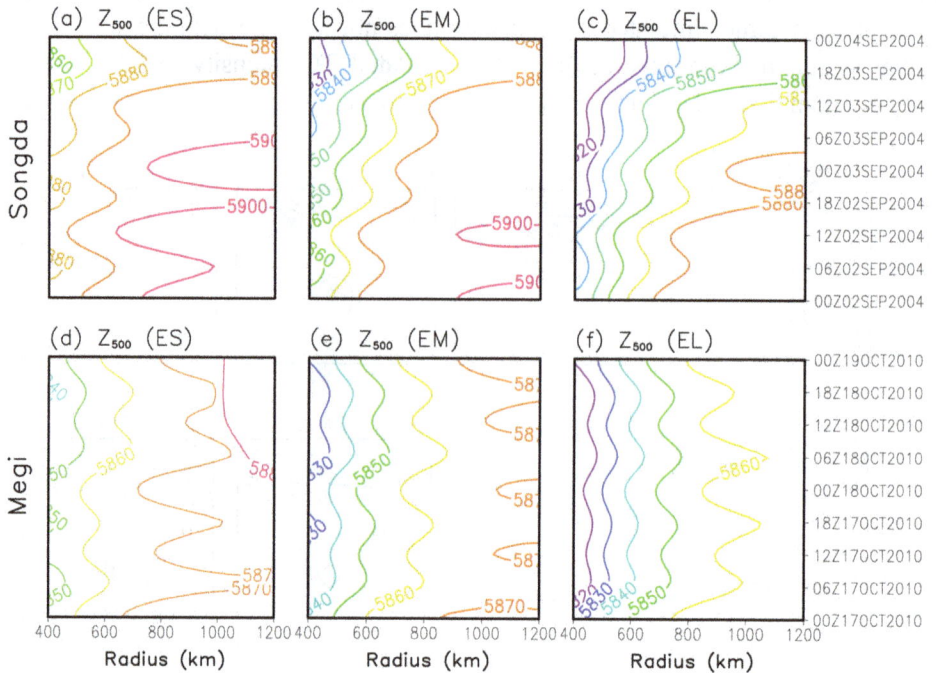

Fig. 6.10. Hovmöller diagram of the azimuthal-averaged geopotential height at 500 hPa (m) in TC outer region before the northward turning of TC in Songda and Megi cases.

In order to investigate this issue, we take a look at the results for the time when there is no significant difference between the sensitivity experiments with different initial storm size. Fig. 6.11 illustrates the radial profiles of geopotential height at 500 hPa at such a time for Songda and Megi cases. Due to the difference in the initial storm size, the simulated geopotential height in the outer region of the

TC varies significantly even after two to three days of integration. The result clearly indicates that the geopotential height in the outer region decreases significantly with the increase in the initial storm size. Although the difference in the 500 hPa geopotential height between the different sensitivity experiments tends to decrease with the increase of the radial distance, remarkable differences occur even at the radial distance of 1200 km. The simulated geopotential height in ES is notably larger than that in EL by up to 21 m in Songda case, and by up to 10 m in Megi case. Meanwhile, the distance between the TC center and the edge of the WPSH is only about 500-1000 km in terms of the geopotential height contour of 5880 m in both Songda and Megi cases (Fig. 6.9). Furthermore, corresponding to increases in the initial TC size, the 500 hPa geopotential height decreases with a higher decreasing rate in areas to the north of the TC than in other areas (see Figs. S6.5 and S6.6). In other words, with the increase in initial storm size, the simulated intensity of WPSH over the region to the north of the TC decreases notably in both Songda and Megi cases.

Fig. 6.11 implies that the pressure gradient at 500 hPa in the TC outer region is closely related to the initial storm size, and increases following the increase of the initial storm size. The pressure gradient below 500 hPa is basically consistent with that at 500 hPa. As suggested by Gopalakrishnan *et al.* [2011] and Sun *et al.* [2013a, 2014b], increases in pressure gradient in the TC outer region could induce increases in inward radial wind speed, and the inflow mass flux (IMF) entering the TC region increases correspondingly. As the TC approaches the WPSH, part of the IMF is transported from the WPSH, contributing to the weakening of the WPSH. To further investigate the possible reasons for the weakening of the WPSH prior to the significant departure of the simulated TC from its realistic location in the EM and EL, we depict the temporal evolution of IMF entering the TC region in the sensitivity experiments for the cases of Songda (2004) and Megi (2010) (Fig. 6.12). It shows clearly that the calculated IMF is sensitive to the initial storm size and increases with the initial storm size, especially during the period prior to the occurrence of significant difference in TC positions between the sensitivity experiments. As the initial storm size increases, more IMF enters the TC region, corresponding to the weakening of the WPSH in the EM and EL (Figs. 6.9 and 6.12). Thereby, the difference in IMF is one of the reasons for the difference in the WPSH simulation between the sensitivity experiments in Songda and Megi cases. Such a decrease in the WPSH intensity leads to a break of the WPSH in the EM and EL. The simulated TC in the EM and EL are subsequently forced to turn northward towards the break in the subtropical ridge. The northward movement of TC will further weaken the intensity of the WPSH near the TC in the EM and EL. This is a positive feedback between the weakening of the WPSH near the TC and northward motion of the TC. However, this positive feedback cannot be initiated in the ES, since the small storm size cannot effectively reduce the geopotential height in the TC outer region. As a result, the intensity of the WPSH over its fringe region on the TC side cannot weaken. It is important to note that, the feedback of the TC on the WPSH intensity is attributed to the larger IMF that enters the TC region rather than the larger contribution of the BEP for the larger TCs (e.g., the TCs in the EM and EL). This is because TCs affect the WPSH through the inflow mass flux, which is not an instantaneous process. The TC motion changes significantly only when the critical change in the WPSH happens (i.e., the break of the WPSH). This also explains why the break of the WPSH is prior to the large departure of the position of the TC center (Fig. 6.9). Additionally, the calculated IMF above 850 hPa is anti-correlated with that below 850 hPa, which offsets the inflow mass fluxes in the lower-troposphere due to the large outflow of mass flux in the upper troposphere. There is a net outflow of mass flux according to the calculated IMF below 100 hPa (Fig. S6.9), which contributes to the intensification and expansion of the TC in both Songda and Megi cases.

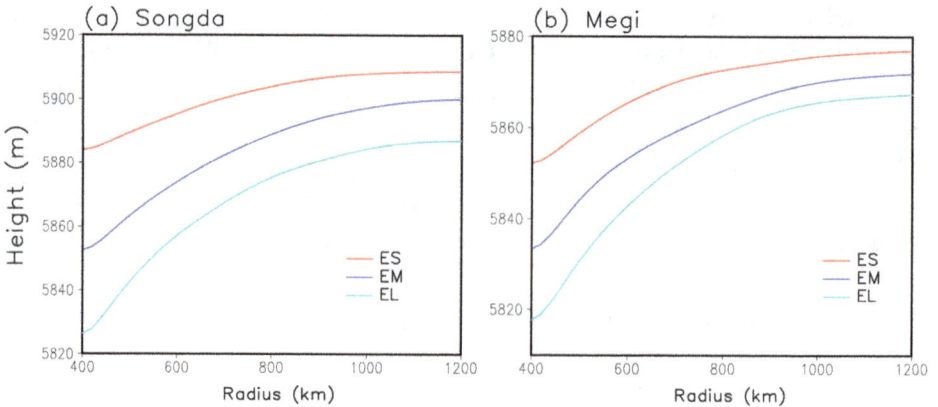

Fig. 6.11. Radial profiles of azimuthal mean geopotential height at 500 hPa in the sensitivity experiments with different initial storm size at 0000 UTC 3 September 2004 for the case study of Songda (2004) and 0000 UTC 18 October 2010 for the case study of Megi (2010), respectively.

Note that, due to the existence of the WPSH in the north of the TC, the difference in geopotential height in the lower-troposphere to the north of the TC between TC outer region and TC inner region is notably larger than that to the other areas. This results in a larger inward pressure gradient force to the north of the TC, which further induces a larger inflow mass flux transporting from the north of the TC outer region to the TC inner region. Due to the larger loss of mass caused by the IMF to the north of the TC, the geopotential height in the north of the TC decreases at a higher rate, which subsequently result in a break of WPSH and thus facilitate the northward turning of the TC in EM and EL experiments.

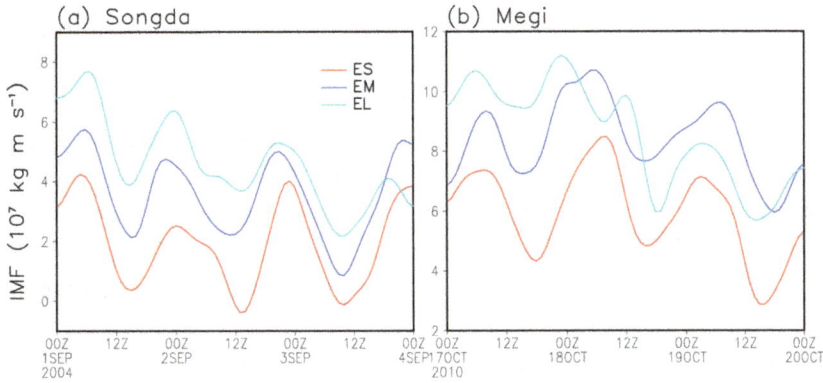

Fig. 6.12. Temporal evolutions of IMF (10^7 kg m s^{-1}) out of a cylinder of radius of 800 km and height of 850 hPa from the TC center in the sensitivity experiments with different initial storm size for the cases of Songda (2004) and Megi (2010).

6.5 Discussions

Consistent with recent studies of idealized TC simulations, the simulated storm size and track is highly sensitive to its initial size with the state-of-art model. It is found that as the initial storm size increases, the TC turns northward earlier and the main body of the WPSH withdraws eastward. Simulations of the ES, EM, and EL for Songda and Megi cases have illustrated how the initial storm size affects the WPSH and TC activities over the WNP. Rather than the BEP, it is the difference in the geopotential height in the TC outer region that is primarily responsible for the large difference in the simulated WPSH and TC track among these sensitivity experiments.

As the initial storm size increases, the inflow mass flux entering the TC region increases, which contributes to the significance decrease in the 500 hPa geopotential height in the TC outer region, especially in areas to the north of the TC. The decrease in the 500 hPa geopotential height is significant even after two to three days of integration. As a result, the simulated intensity of WPSH over its fringe to the north of the TC decreases notably as the TC approaches the WPSH and the fringe of the WPSH is within the TC outer region. Such a decrease in the WPSH intensity leads to a break of the WPSH in the EM and EL simulations. The simulated TCs in EM and EL are forced to turn northward towards the break in the subtropical ridge. The northward motion of the TC will further weaken the intensity of the WPSH in the EM and EL simulations. This is a positive feedback between the weakening of WPSH near the TC and the northward moving of the TC, which contributes to the large difference in the WPSH intensity and the TC motion simulated by the sensitivity experiments for both Songda and Megi cases.

The simulated storm intensities are not shown in this study since the grid spacing of 20 km is too coarse to accurately reproduce the storm intensity. For the reason of coarse resolution, most RCM studies are carried out to investigate the TC frequency and motion instead of the TC intensity. Moreover, the simulated TC intensity is not crucial in understanding how the initial storm size influences the TC motion and WPSH intensity. As suggested by Holland and Merrill [1984] and Stern and Nolan [2009], the geopotential height and pressure in the TC outer region are mainly determined by the TC size, while no relationship is found between the TC size and TC intensity. Thus, although the 500 hPa geopotential height in the TC outer region decreases notably with the increase in the initial storm size, the simulated TC intensity does not increase correspondingly. However, note that the simulated TC intensity in the EM is stronger than that in the EL for both Songda and Megi cases. This also indicates that the change in the WPSH intensity is more related to the TC size than to the TC intensity. Namely, it is the larger TC rather than the stronger TC that leads to the larger decrease in 500 hPa geopotential height and thus the weakening of the WPSH.

It is also found that HA plays the most important role in determining the differences in TC motion between the EL and EM by accelerating the northward movement of the TC in the EL. The impacts of VT and DH on TC motion are less significant than that of HA. Thereby, it is the difference in HA that is responsible for the difference in simulations of TC motion between the sensitivity experiments. HA includes not only the contribution of environmental flow, but also the contribution of the beta-effect propagation (BEP). However, results of the calculated meridional BEP speed shows that, compared with the contribution of HA to TC motion, the BEP contributes little to the TC motion. Further analysis shows that, it is the difference in the geopotential height in the TC outer region that plays a critical role in determining the difference in the WPSH and thus TC motion between these sensitivity experiments.

In this chapter, we mainly focus on the dynamic interaction between the TC and WPSH. This does not mean that there is no thermodynamic interaction between them. To further explore the source of the difference in the simulated WPSH, we have also compared contributions of each individual term to temperature tendency. It is found that for Megi case, both dynamic and thermodynamic processes contribute to the withdrawal of the WPSH, and the former makes a greater contribution than the latter. However, for Songda case, due to insufficient atmospheric moisture in the lower-troposphere, DH (i.e. the thermodynamic process) contributes little to changes in the WPSH, while HA (i.e. dynamic process) plays a major role in determining the difference in the WPSH simulation between the sensitivity experiments. This indicates that for Megi case, the dynamic feedback of TC alone can change the WPSH significantly and leads to unrealistic weakening of the WPSH and the early recurvature of the TC. Therefore, despite the thermodynamic process contributing to the difference in the WPSH, it is the dynamic process that is responsible for the difference in the WPSH simulations between the sensitivity experiments for both Songda and Megi cases.

It is important to note that, the physical mechanism revealed here cannot explain all the observational TC tracks. As suggested by Lee *et al.* [2010], most of the 18 (16) persistently large (small) TCs from 145 total TCs move northwestward (westward), which can be explained by the mechanism disclosed in this study. However, our mechanisms may not be effective to explain the tracks of the rest TCs which are neither persistently large nor persistently small. In fact, besides the storm size, the large-scale variability of WPSH and the position of TC genesis relative to the WPSH are also key factors in influencing the TC track. Thereby, beside the storm size, more factors should be considered to provide a rational explanation for the most of the observational TC tracks.

The physical mechanism not only helps us better understand the feedback of TC on the WPSH, but also emphasizes the importance of a correct representation of the initial storm size for realistic simulation of the WPSH and TC track. Incorrect representation of the initial storm size could be a reason for errors in RCM studies, although this requires extra model and case studies to confirm. The findings have shed light on the role of initial storm size in the simulation of the WPSH and TC, which might be helpful for resolving the difficult problems in operational forecast of the WPSH and TC track. It is noteworthy that, in addition to the initial storm size, the model-simulated storm size is sensitive to model physics (e.g., SST, environmental humidity, etc), this will be a topic for future research.

Chapter 7

Model Convergence in Simulations of Tropical Cyclone at Grey-Zone Resolutions

7.1 Introduction

Current NWPs use high-resolution models with horizontal grid spacing of 1-10 km [Lean *et al.*, 2008; Roberts and Lean, 2008]. For such high-resolution models, further refinements of both physical parameterizations and numerical techniques are required [Steppeler *et al.*, 2003]. One of the major problems in NWP from a physics perspective is the treatment of deep moist convection. Due to the assumptions and the closure hypotheses upon which cumulus convective parameterization is based, many CPSs may not be suitable for the NWP models with a grid spacing of a few kilometers [Molinari and Dudek, 1992; Hammarstrand, 1998]. It has been debated whether CPS is necessary for numerical predictions of convective systems at those grid spacings. Gerard [2007] used the term "grey-zone resolution" to refer to this type of resolution. Yu and Lee [2010] suggested that this type of resolution can range from 1 km to 5 km. Generally, the grid spacings below the grey-zone resolution are considered sufficient for explicit simulation of convective weathers, and thus CPS is not necessary [Liu *et al.*, 1997]. In contrast, many studies showed that there will be some problems in high-resolution simulations without CPS such as irresolvable convection and underprediction or overprediction of precipitation and thus use of CPS is also recommended for higher resolution simulations [Kotroni and Lagouvardos, 2004; Deng and Stauffer, 2006]. Without using CPS in high-resolution runs (e.g., 2.8- or 1-km grid spacing runs), the model cannot properly reproduce some convective processes in evolving cumulus clouds [Niemelä and Fortelius, 2005; Craig and Dörnbrack, 2008]. Similar problems could also be encountered in TC simulations. Rogers *et al.* [2006] cited inadequate computational resources to run operational models at sufficiently high spatial resolution, along with incomplete representation of important physical processes, as two reasons for the slow improvement in TC intensity prediction. Without improvements in representation of model physics, the decrease of grid spacing alone could not significantly and continuously improve TC intensity forecast [Fierro *et al.*, 2009]. Mass *et al.* [2002] stated that "decreasing grid spacing in mesoscale models to less than 10-15 km generally improves the realism of the results but does not necessarily significantly improve the objectively scored accuracy of the forecasts". Gentry and Lackmann [2010] (hereafter, GL10) suggested that, as grid spacing is decreased, the structure and evolution of Hurricane Ivan (2004) undergo significant changes, with the finest resolution, the 1-km resolution, exhibiting a markedly different structure from those of the other simulations. It is assumed that perfect models should produce convergent TC intensity and structure as the grid spacing decreases although the converged solution does not necessarily approach the truth; otherwise, the model is considered not convergent. The GL10 results indicate that the model solution has not yet converged at the grid spacings ranging from 8 km to 1 km. Bryan *et al.* [2003] also found that the model didn't converge with grid spacings well below 1 km in their convective system simulation. In fact, it is impossible for the simulated storm to intensify indefinitely as the grid spacing decreases, i.e., the model will for sure converge at a certain very high resolution, but this resolution may not be practical to run with under the present-day computer resources. Moreover, it is still a central question, as addressed by Kain *et al.*[2008], whether the extra computer resources required to run high-resolution models produce a worthwhile increase in forecast accuracy.

Why is there an absence of model convergence in high-resolution simulations? It might be expected that the magnitude of vertical motions would increase as smaller grid spacing is used, as updrafts are more adequately resolved. However, the conclusions of Bryan *et al.* [2003] suggested that higher resolution could eventually weaken the magnitude of vertical motions as detrimental processes such as entrainment with grid-scale turbulence begin to be resolved. GL10 focused on what changes occur in the representation of physical processes important to TC intensity as grid spacing decreases, and implied that these changes may contribute to the additional and unnecessary strengthening of the simulated TC at high resolution.

Furthermore, CPSs are not generally designed for the finer grid lengths utilized in this chapter, nor in the inner-core regions of TCs [Molinari and Dudek, 1992; GL10]. The use of CPS would implicitly account for a portion of the eyewall updraft, thereby weakening the upward branch of the secondary circulation and the compensating grid-scale subsidence within the eye [GL10]. This problem is important because many studies on convection organized on the mesoscale (including TCs) systematically make use of horizontal grid spacing of less than 10 km within the innermost mesh. Also, the future real-time hurricane forecasts using mesoscale models will increasingly make use of cloud-resolving grid spacing (i.e., grid spacing < 5 km) as computer power increases. Thus, it is urgent to find a reasonable CPS suitable for the high-resolution model to obtain more realistic results with a stronger model convergence and thus a better forecast value.

Despite some studies concerning the weak model convergence in simulation of TC intensity as grid spacing is decreased, they did not provide detailed and systematic analysis on the reasons for the weak convergence, much less the impact of CPSs on this model convergence. Even it is not clear yet whether CPS could have a remarkable impact on this model convergence. In this chapter, through a series of sensitivity experiments, the possible reasons for the weak model convergence in simulating TC intensity at various grey-zone model resolutions, the impacts of different CPSs on this convergence, and whether it is possible to find a reasonable CPS that works better at high resolution to make this convergence stronger will be addressed. The results obtained herein could have potential applications to widely used numerical research and forecast models.

7.2 Experimental design

Typhoon Shanshan (2006) is selected as a case for this study. It was generated over the western Pacific (16.7°N, 134.9°E) at 1200 UTC 10 September 2006. After its occurrence, it moved westward with its intensity increased unceasingly. At 1800 UTC 14 September, it moved to the east of Taiwan (20.7°N, 124.6°E), and developed into an extremely severe typhoon. Then, it turned northeastward, and made a landfall in Japan at about 1000 UTC 17 September. Shanshan has the typical turning track over the sea with the characteristics of high intensity, long duration, and fast development, which are illustrated by the best-track data obtained from JTWC.

The model used here is also the Advanced Research Weather Research and Forecasting (WRF-ARW) model. The model domain is triply-nested (DM1, DM2, and DM3) for most cases, with an added nest (DM4) for 1-km resolution runs. The inner meshes (DM2, DM3, and DM4) automatically move to follow the model storm [Skamarock *et al.*, 2008]. The initial and lateral boundary conditions were obtained from the 1°×1° NCEP final analysis data (FNL) at 6-h intervals. The TC Bogus scheme in the WRF model was adopted to build the initial field [Skamarock *et al.*, 2008]. In order to consider the effect of the cool upwelling on TC intensity [Chan *et al.*, 2001; Zhong and Zhang, 2006], daily SST was updated using the TRMM Microwave Imager (TMI) level-1 standard product at a 0.25° × 0.25° resolution, provided by the Earth Observation Research Center (EORC)/National Space Development Agency (NASDA) (http://www.eorc.nasda. go.jp/TRMM).

The model integration started at 0000 UTC 14 September 2006 and ended at 1200 UTC 16 September 2006, with a total of 60 h for both the intensification period and the steady-state period of Shanshan. The Yonsei University non-local-K planetary boundary layer scheme [Hong *et al.*, 2006] and Monin-Obukhov surface layer scheme [Paulson 1970; Dyer and Hicks 1970; Webb 1970; Beljaars 1994] are used in this study. It should be mentioned that the present study uses the hybrid (double moment for ice crystals and single moment for all other species) Thompson *et al.* [2004] microphysical scheme, which is a new scheme with ice, snow and graupel processes suitable for high-resolution simulations. In the Thompson scheme, riming growth of snow is required to exceed depositional growth of snow by a factor of 3 before rimed snow transfers into the graupel category, resulting in more realistic values for graupel mixing ratio in the eyewall [Fierro *et al.*, 2009].

There are a total of four groups of simulations: simulation with (1) no CPS (NOCP), (2) Betts-Miller-Janjić CPS (BMJ) [Betts, 1986; Betts and Miller, 1986; Janjić, 1994; Janjić, 2000], (3) Kain-Fritsch CPS (KFEX) [Kain and Fritsch, 1990; Kain, 2004], and (4) Grell 3D ensemble CPS (GR3D) [Grell and Devenyi, 2002; Skamarock *et al.*, 2008] in the finest mesh. Each group contains four simulations at 1-, 3-, 5-, 7.5-km horizontal resolutions in its finest mesh. For example, GR3D1, GR3D3, GR3D5, and GR3D7.5 represent the GR3D simulations with grid sizes of 1, 3, 5 and 7.5 km, respectively. In addition, the GR3D scheme was first introduced in WRFV3.0, and so is new, and not yet well tested in many situations. Based on an ensemble mean approach, it shares a lot in common with the Grell and Devenyi [2002] scheme, but the quasi-equilibrium approach is no longer included among the ensemble members. The scheme is distinguished from other cumulus schemes by allowing subsidence effects to spread to neighboring grid columns, making the scheme more suitable to grid spacing less than 10 km [Skamarock *et al.*, 2008]. Thus, in the following sections, it is worthy of more attention to its performance in the higher resolution runs.

We adopted a reasonable approach to maintain the consistency of model design and thus effectively exhibited the impact of horizontal resolution and CPS in a nested domain configuration: The DM1, DM2 and DM3 in all cases cover the same areas with horizontal dimensions of 2475×3375, 1500×1500 and 785×785 km^2, respectively (Table 7.1), and the choice made in the physics and dynamical setup except for the CP in the innermost domain is also consistent with each other cases. In addition, the DM3 and DM4 in 45-15-5-1 cases used the same CP. In this chapter, we wish to put emphasis on the need to keep consistency between all cases. Also, the analysis presented herein entirely focuses on the innermost-domain data, whose dimensions were chosen to cover the great majority of the TC's convection in all cases.

Table 7.1. Dimensions of the domains for the four groups of simulations with different horizontal resolutions in their innermost mesh.

Case	Grid spacing	DM1	DM2	DM3	DM4
7.5 km	45-15-7.5 km	55 × 75	100 × 100	105 × 105	none
5 km	45-15-5 km	55 × 75	100 × 100	157 × 157	none
3 km	45-15-3 km	55 × 75	100 × 100	261 × 261	none
1 km	45-15-5-1 km	55 × 75	100 × 100	157 × 157	501 × 501

7.3 Sensitivity of TC activity to horizontal resolution and CPS

7.3.1 *Storm track*

Fig. 7.1 compares the storm tracks from these sensitivity experiments overlaid with the JTWC best track. It is important that WRF-ARW was able to capture the track of Shanshan well, and in all cases the simulated storms followed a similar track to that of the JTWC, except for the 7.5-km runs which translated a little more rapidly than the observed in the last 24-h simulation. Including those in 7.5-km runs, the simulated time-averaged track errors are no more than 100 km. This is not a surprising result and can be attributed to the consistency of model design for DM1 and DM2 in all cases: As Marks and Shay [1998] suggested that, unlike storm intensity, track prediction depends more on larger-scale processes that can be resolved with coarse grid prediction models. Namely, in all cases, the similarity of simulated tracks results from the similar larger-scale environmental fields produced by the DM1 and DM2 simulation. On the other hand, the similar simulated tracks, in turn, further ensure the similar larger-scale environmental fields. Therefore, due to the consistency in model design (including domain sizes and the choice made in the physics and dynamical setup) and thus TC tracks and larger-scale processes, it is more reliable to chapter the impact of horizontal resolution and CPS on storm intensity and structure without interference from other factors.

A further comparison of the simulated tracks among different resolution runs reveals that the track error decreases notably as the horizontal resolution of the innermost domain decreases from 7.5 km to 5 km. This is because the storm track depends on not only larger-scale steering flow but also storm structure [Fiorino and Elsberry, 1989], and the location of storm center will be more accurate due to the more realistic storm structure produced in higher-resolution runs. In terms of the storm center position, the higher-resolution runs (i.e. 3-km and 1-km runs) in the four CPS groups all presented well. However, the simulated track error changes little when the resolution further decreases from 5 km to 1 km. In addition, in the fine-mesh runs (resolution < 5 km), the choice of CPS had little effect on storm track; the turning of the track from northwestward to northeastward was captured decently by all the four CPS groups of runs. Thus, in the fine-resolution cases (such as resolution < 5 km), the decrease of resolution and the choice of CPS in the WRF-ARW cannot contribute to the notable improvement in simulation of TC track.

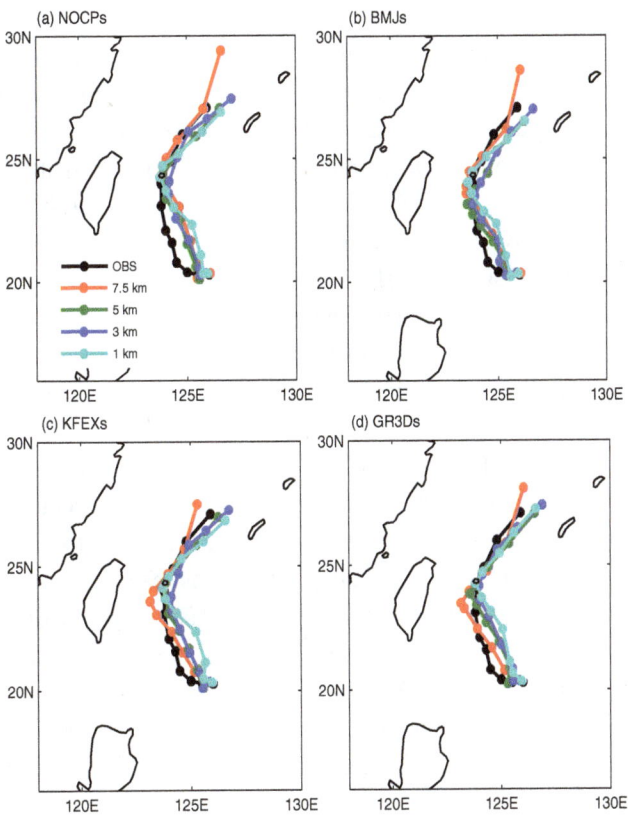

Fig. 7.1. Storm tracks overlaid with the observed best track (black disks) at 6-h intervals for the sensitivity experiments at various resolutions (a) NOCP, (b) BMJ, (c) KFEX, (d) GR3D.

7.3.2 *Storm intensity*

Though some recent studies have noticed the impact of grid spacing on TC intensity at grid spacings below 5 km, no final conclusion has yet been reached on this matter. Fierro *et al.* [2009] proposed that the features in finer-resolution simulations that tend to weaken TCs (i.e., smaller area of high surface fluxes and weaker total updraft mass flux) compensate for the features that favor TC intensity (i.e., weaker eyewall asymmetries and larger radial gradients); thus, this will result in a similar intensity between finer- and coarser-resolution runs. However, GL10 suggested that, due to changes in the representation of physical processes important to TC intensity, the simulated TC at higher resolution could be additionally strengthened. In this chapter, our results are basically consistent with GL10 in that the finer-resolution cases do exhibit stronger TC intensity than those at coarser resolution, and the model solution has not yet converged at these grid spacings.

To demonstrate the impact of grid spacing and CPS on TC intensity, we compare in Figs. 7.2 and 7.3 the time series of minimum sea level pressure and maximum wind speed (MWS) at 10 m for the sensitivity experiments and the JTWC best track. Davis *et al.* [2008] found that, TC intensity did show notable changes as grid spacing was decreased from 4 km to 1.3 km (about 20 hPa in MSLP and 13 m s^{-1} in MWS). The similar result was also obtained in GL10. In our simulations, except for the BMJ cases, the other three groups of cases basically follow this principle: TC intensity underwent a relatively small change as the grid spacing decreases from 5 km to 3 km, while a significant increase was observed from 3 km to 1 km. However, in BMJ cases, TC intensity increases slightly as the grid spacing decreases from 7.5 km to 5 km but an extreme increase is observed from 5 km to 1 km. In addition, except for BMJ cases, due to the enhanced convection at grid-scale, some deepening of the TC does occur as the grid length decreases from 7.5 km to 5 km, which is consistent with the results of GL10 for grid lengths from 8 km to 6 km. Thus, in all cases, as grid spacing is decreased to below 3 km, more deepening per decrease in grid spacing is realized. This indicates that based on the present CPS, the model solution has not yet converged, especially with grid spacing below 3 km. Moreover, though the model convergence in GR3Ds is relatively strong and better than in NOCPs, BMJs, and KFEXs, the simulated peak intensity in its 1-km run is still about 10 hPa and 5 m s^{-1} stronger than that in its 3-km run.

Overall, the increase in storm intensity is not uniform across similar reductions in grid spacing, and there is a significant increase in intensity as the grid spacing is decreased from 3 km to 1 km. On the other hand, due to the significant impact of CPS on TC intensity and thus the model convergence, we expect to find a suitable and reasonable CPS to tackle the problem on the weak model convergence.

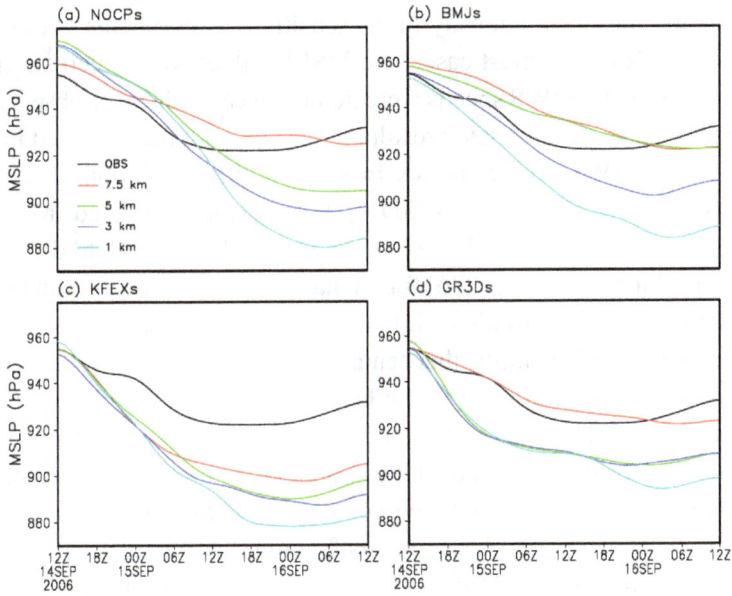

Fig. 7.2. Temporal evolutions of MSLP overlaid with the JTWC best track (black line) for the sensitivity experiments (a) NOCP, (b) BMJ, (c) KFEX, and (d) GR3D.

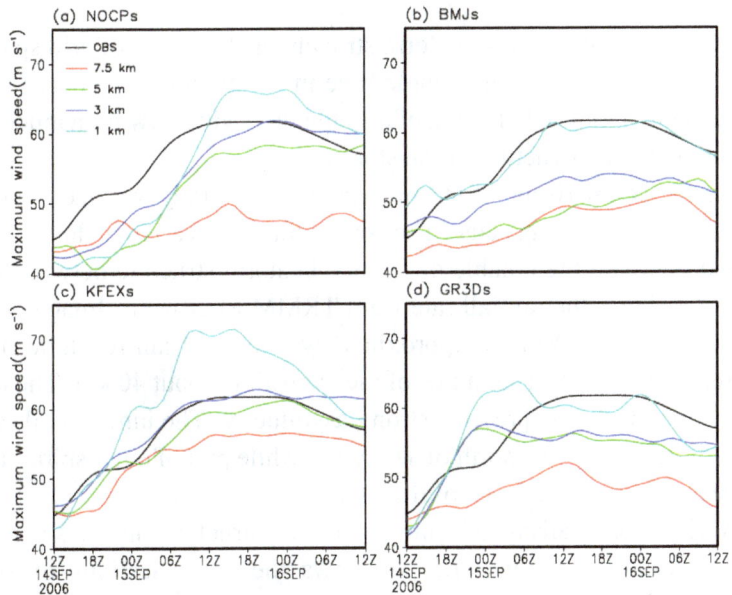

Fig. 7.3. Temporal evolutions of maximum 10-m wind speed overlaid with the JTWC best track (black line) for the sensitivity experiments (a) NOCP, (b) BMJ, (c) KFEX, and (d) GR3D.

Comparing the storm intensity in the sensitivity experiments with that in the best analysis finds that in most cases, the MSLP values were much stronger than the observed while the MWS speeds were notably weaker than the observed. Thus, it is difficult to make the simulated results consistent with the observed in terms of both MSLP and MWS. For example, in some higher-resolution cases (such as NOCP3, BMJ1, KFEX3, and GR3D1), though the simulated MWSs were relatively closer to the observed, the simulated MSLPs were quite different from the observed. Zhu and Zhang [2006] found that the maximum wind can be affected by localized convective activities while the minimum sea-level pressure tends to be a system-integrated quantity that tends to be a more reliable measure of the vortex intensity. In general, comparing GR3Ds with the other three CPS groups reveals that the evolutions of MSLP and MWS in GR3Ds were more similar to those observed. However, as a new CPS designed for high-resolution simulation, GR3D still has some problems to be solved. The simulated storms in GR3Ds presented a much earlier and faster intensification than that observed during the development period (about 40 hPa and 10 m s^{-1} from 1200 UTC 14 September to 0600 UTC 15 September).

7.3.3 *Storm structure*

In performing the comparisons of storm structure as functions of grid spacing and CPS, we focus our analysis on a single time in the mature stage here (e.g., 0000 UTC 16 September 2006). Selection of a single time will allow for a much detailed examination of spatial structure of the storm.

Before the comparisons of storm structure, we firstly analyze the observed storm structure at the mature stage. The lack of intensity variation during this stage allows us to highlight the notable difference in storm structure between different cases. Fig. 7.4 shows the rainfall rate from TRMM Microwave Image (TMI) and Precipitation Radar (PR) (at an approximately 4 km × 4 km resolution) at 0431 UTC 16 September 2006. The radius of the eyewall is about 40 km. The observed inner-core structure of TC presents strong asymmetry. The inner spiral rainbands are mainly distributed in the north of TC center while part of outer spiral rainbands appears west, southwest and east of the TC.

To verify the realism of simulated TC structure in those sensitivity experiments, we plotted Figs. 7.5. It shows the cross sections of simulated instantaneous rainfall rate at the mature stage of TC Shanshan. Compared to the observation in Fig. 7.4, the simulated rainfall over the TC eyewall is notably larger than the observed. It needs to be stated here that relative to the magnitude of the simulated rainfall, we are more concerned with the spatial distribution of the TC rainfall. Moreover, following the decrease of grid spacing, the simulated rainfall rate did not get closer to the observed, but even became worse in NOCPs, BMJs, and KFEXs because the radii of eyewall and the asymmetric feature in their 1-km runs were quite different from the observed. Compared with the other three groups of CPS cases, GR3Ds reproduced a relatively realistic spatial distribution of spiral rainbands, especially in the higher-resolution runs (e.g., GR3D1) in terms of the radius and location of spiral rainbands. Furthermore, the radii of the eyewall decrease notably as the grid lengths decrease from 7.5 km to 1 km (especially from 3 km to 1 km) in those four groups of CPS experiments. This may be responsible for the difference in TC intensity between the coarser-resolution run and higher-resolution run, which will be further discussed in the next section.

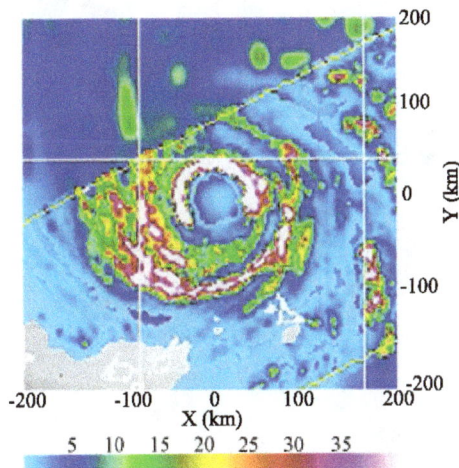

Fig. 7.4. Rainfall rate (mm h^{-1}) from TMI/PR at 0431 UTC 16 September 2006.

7.4 Possible reasons for the weak model convergence

7.4.1 Energy and mass exchange

7.4.1.1 *Energy exchange at air-sea interface*

Previous studies suggested that TCs intensify and maintain themselves against surface frictional dissipation by extracting energy from the underlying oceans. Thus, energy exchange at the air-sea interface is the key to the intensity change of a TC [Malkus and Riehl, 1960; Black and Holland, 1995]. The wind-induced surface heat exchange, which describes a positive feedback between the increase in surface energy flux (SEF) and the surface wind speed in the near-core region of a TC, is viewed as the dominant process that controls the rapid intensification of a TC [Emanuel, 1986; Rotunno and Emanuel, 1987]. SEF is the sum of sensible-heat (SH) and latent-heat fluxes (LH). As suggested by Sun *et al.* [2013b], SH is dependent on the surface wind speed and air-sea temperature difference (ASTD), which can be defined as SST minus air temperature at 2 m, while LH is dependent on the surface wind speed and air-sea moisture difference (ASMD), which can be defined as the saturation specific humidity minus air specific humidity at 2 m.

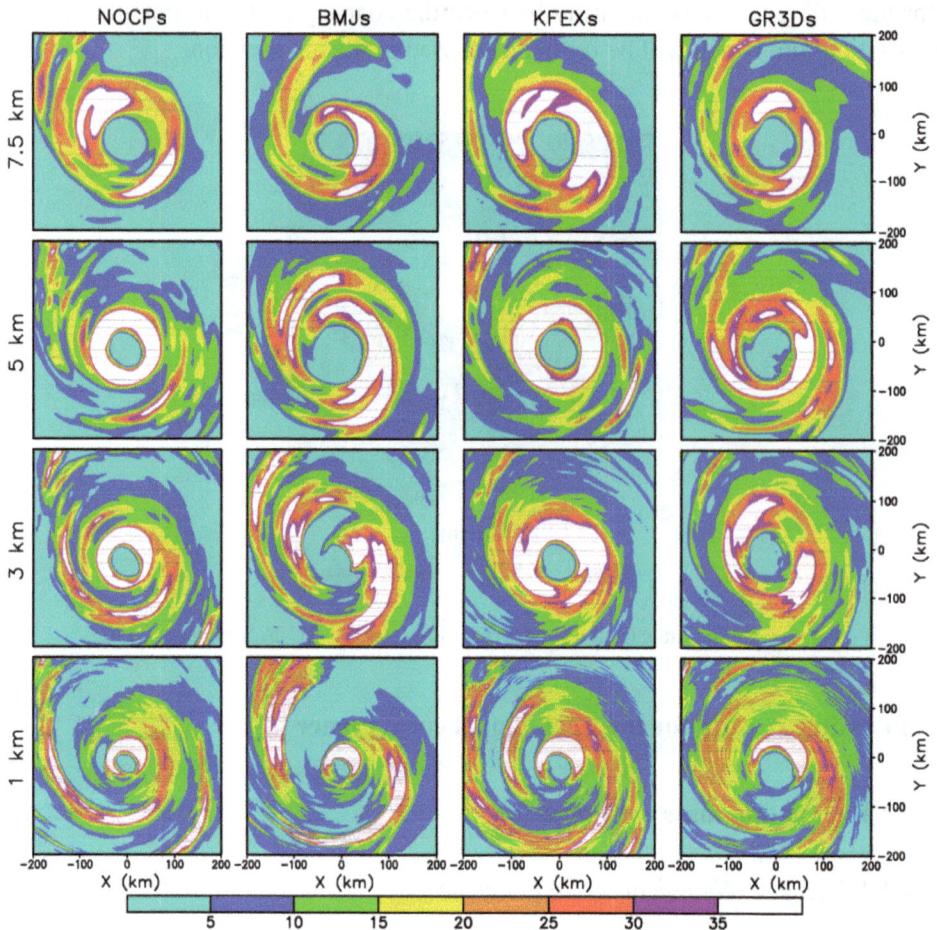

Fig. 7.5. Horizontal cross sections of instantaneous rainfall rate (mm h^{-1}) from model runs (including cumulus precipitation and grid scale precipitation) at 0400 UTC 16 September 2006.

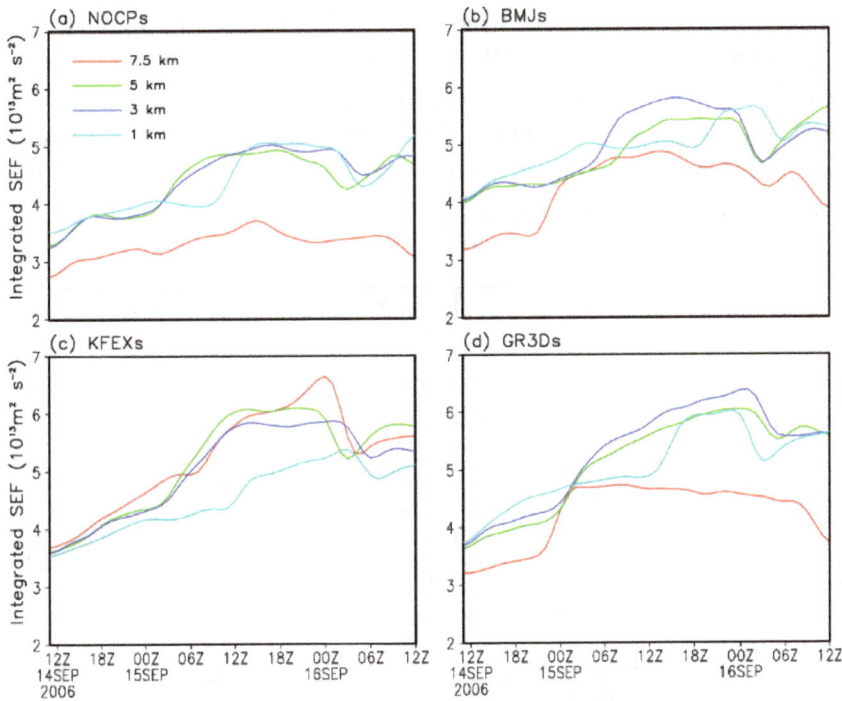

Fig. 7.6. Temporal evolutions of area-integrated surface energy flux (10^{13} m^2 s^{-2}) within a radius of 200 km for the sensitivity experiments (a) NOCP, (b) BMJ, (c) KFEX, and (d) GR3D.

In order to find the possible reasons for the weak convergence of the WRF-ARW in simulation of storm intensity mentioned above, area-integrated kinetic energy and SEF were calculated. Fig. 7.6 shows the temporal evolutions of area-integrated SEF within a 200-km radius. Owing to the effect of abnormal warm SST and other favorable conditions, the area-integrated SEF increase continuously from 1200 UTC 14 September to 0300 UTC 16 September. Note that a significant reduction process of SEF occurred at about 0300 UTC 16 September 2006 in these experiments, for the simulated TC passed over a cold pool near Miyako Island during that period. In our experiments, the area-integrated SEFs presented an increasing trend as the horizontal spacing decreased from 7.5 km to 5 km except for KFEXs. The increased SEF in BMJ5 and GR3D5 could be attributed to its stronger area-averaged surface wind speeds, while the increased SEF in NOCP5 could be attributed to not only its higher surface wind speed but also its higher ASTD and ASMD. However, there was no significant difference between the integrated SEF in 5-km runs and 3-km runs in all the four groups of CPS experiments. As the grid spacing further decreased from 3 km to 1 km, the integrated SEF in 1-km runs was not notably higher and even weaker because of the weaker area-averaged surface wind speeds, which is a result from the smaller radius of the maximum wind (RMW) and thus the smaller extent of strong wind region in 1-km runs. Thereby, the view from integrated SEF might explain the difference in storm intensity between 7.5-km runs and 5-km runs, but could not explain the difference between 3-km runs and 1-km runs, and thus other factors must be promoting storm intensity in 1-km runs.

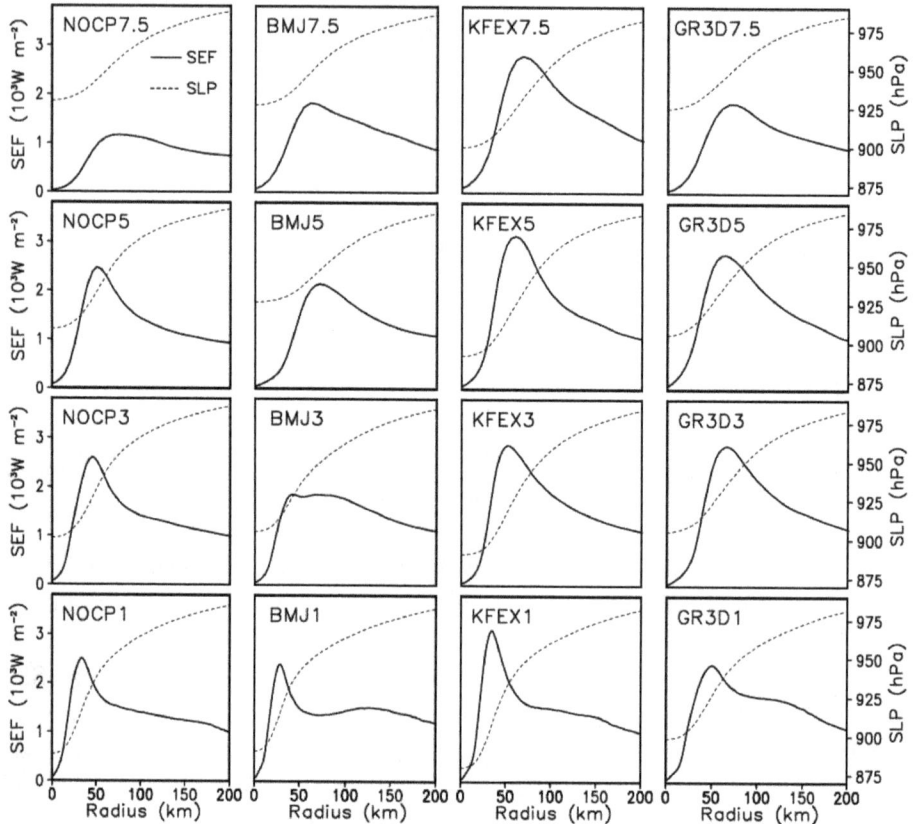

Fig. 7.7. Radial distributions of the azimuthal mean surface energy flux (SEF) and sea level pressure (SLP) averaged in the TC mature stage from 1800 UTC 15 September to 0600 UTC 16 September.

The radial distribution of the azimuthal mean SEF and sea level pressure (SLP) averaged in the TC's mature stage are shown in Fig. 7.7. Similar to the difference in storm intensity between different resolution runs, the simulated SEF (SLP) in 5-km runs are notably greater (lower) than those in 7.5-km runs, especially in NOCPs and GR3Ds. As the grid spacing further decreased from 5 km to 3 km, except for BMJs, there were no substantial differences between 5-km run and 3-km run in terms of the SEF and SLP. Moreover, following the grid spacing decrease from 3 km to 1 km, except for GR3Ds, the maximum values of azimuthal mean SEF in the other three CPS groups of experiments never underwent a significant change, but their radial positions shifted inward notably and the values of SEF outside the eyewall decreased substantially. Most importantly, this will lead to sharper radial gradients of SEF at the inner edge of the eyewall and thus sharper SLP gradients, which will bring about a more intense storm [Wang, 2009]. However, in GR3D1, although the maximum value of SEF also shifted inward, the SEF outside the eyewall did not decrease remarkably. This could not lead to a notable increase in the radial SEF and SLP gradients near the eyewall, neither to the increase of the simulated storm intensity in GR3D1.

Above all, from the viewpoint of energy exchange at the air-sea interface, it is inferred that the sharp radial gradient of SEF near eyewall may be an important reason for the weak model convergence in storm intensity between 3-km runs and 1-km runs in NOCPs, BMJs, and KFEXs. As a new CPS which is more suitable to high-resolution models, GR3D presents a relatively better model convergence in the simulation of storm intensity due to smaller radial gradients of SEF and SLP.

7.4.1.2 *Inflow and vertical mass fluxes in the boundary layer*

Drawing on the works by Hendricks *et al.* [2004] and Montgomery *et al.* [2006], a secondary vortex enhancement mechanism was identified and demonstrated in Tory *et al.* [2006]. System-scale intensification was found related to the large-scale response to the net vertical mass flux driven by diabatic heating in the convective cores. This led to the enhancement of the secondary circulation in a manner akin to the classical Eliassen model of a balanced vortex driven by heat sources.

The averaged inflow and vertical mass fluxes, and integrated boundary layer (BL) kinetic energy are listed in Table 7.2. As the grid spacing decreased from 7.5 km to 5 km, the strength of the secondary circulation, measured by the inflow and vertical mass fluxes, showed a notable intensifying in nearly all the four groups of CPS experiments (except for the inflow mass flux in KEFXs), consistent with the larger area-integrated SEF (Fig. 7.6), larger hydrometeor mass aloft, and stronger updraft speeds presented in section 7.4.2.3 (Figs. 7.12 and 7.13). A similar trend was simulated for the BL kinetic energy except for that in KFEXs, which also indicted the stronger near-surface wind fields. Although the BL kinetic energy in the 3-km runs is slightly higher than that in 5-km runs, the strength of the secondary circulation showed a weakening trend as the grid spacing decreased from 5 km to 3 km. However, compared with those in 3-km runs, the simulated storms in 1-km runs produced much smaller low-level inward mass flux, which by virtue of mass conservation must be consistent with smaller vertical mass flux in the eyewall and contributed to a smaller BL kinetic energy. This shows that, as the grid spacing is less than 5 km, the secondary circulation is weaker at finer resolution, which is consistent with smaller area-integrated SEF (Fig. 7.6), and unfavorable for storm intensity. As that in area-integrated SEF, the smaller integrated mass flux and BL kinetic energy in higher-resolution runs may be a result from the smaller RMW and thus the smaller extent of strong wind region in 1-km runs (shown in section 7.4.2.3). However, it appears that at finer resolution, these unfavorable conditions for storm intensity (e.g., the smaller area-integrated SEF and mass flux), could be counteracted by other favorable conditions (e.g., larger radial gradient and other factors), and result in a stronger storm, which will be discussed later in the text.

Table 7.2. Time-averaged inflow (2nd column) and vertical mass flux (kg m^{-1}) (3rd column) out of a cylinder of radius of 100 km and height of 1 km MSL from the storm's center in the TC mature stage. The rightmost column shows the time-averaged BL integrated kinetic energy (m^2 s^{-2}) within a cylinder of height of 1-km above MSL and a radius of 200 km from the storm's center in the TC mature stage from 1800 UTC 15 September to 0600 UTC 16 September.

Cases	Inflow mass flux ($\times 10^9$ kg m s^{-1})	Vertical mass flux ($\times 10^9$ kg m s^{-1})	BL kinetic energy ($\times 10^{13}$ m^2 s^{-2})
NOCP7.5	-8.76	7.35	4.52
NOCP5	-9.26	13.19	4.81
NOCP3	-8.83	10.57	4.97
NOCP1	-7.20	7.32	4.94
BMJ7.5	-8.14	7.37	4.40
BMJ5	-12.18	8.60	5.10
BMJ3	-11.25	7.19	5.24
BMJ1	-5.08	5.31	4.95
KFEX7.5	-16.81	12.62	6.37
KFEX5	-14.21	16.32	6.17
KFEX3	-14.31	13.76	6.19
KFEX1	-7.76	8.21	5.46
GR3D7.5	-10.20	8.97	4.68
GR3D5	-13.72	12.58	5.50
GR3D3	-14.28	11.59	5.80
GR3D1	-8.32	7.39	5.75

7.4.2 *Differences in the storm structures*

7.4.2.1 *Asymmetric structure*

Fierro *et al.* [2009] showed that during the steady-state period of a TC's development, the coarser-resolution cases exhibited larger-amplitude eyewall asymmetries, which has been shown to act as a potential brake on storm intensity increase [Peng *et al.*, 1999; Yang *et al.*, 2007]. To quantify the amplitude of the eyewall asymmetries in our sensitivity experiments, the simulated 10-m wind speed (in plane Cartesian coordinate) is first interpolated onto polar coordinate grids. The projected pole is taken as the TC center (the point with minimum wind speed). Make a Fast Fourier transform (FFT) to the polar coordinate grid of the wind speed during the mature stage of the TC, and then separate the wind into symmetric flow and perturbation flow that correspond to each azimuthal wavenumber, similar to what was done in Sun *et al.* [2012]. In addition, consistent with Fierro *et al.* [2009], there is almost no difference between the results of FFT analysis on wind speed and that on radar reflectivity; thereby the eyewall asymmetries could be represented by the asymmetric structure of wind speed.

Table 7.3. Amplitude of Fourier spectral decomposition of wind speed at 10 m for 3-km and 1-km runs averaged in the TC mature stage. All values are in m s^{-1}.

Cases	NOCP3/NOCP1	BMJ3/BMJ1	KFEX3/KFEX1	GR3D3/GR3D1
Symmetric flow	52.00 / 51.85	45.36 / 50.76	54.93 / 54.81	50.11 / 47.90
Wavenumber 1	19.67 / 11.36	8.88 / 7.17	14.14 / 9.73	5.21 / 11.84
Wavenumber 2	5.81 / 3.60	2.91 / 4.36	3.91 / 2.34	4.14 / 4.47
Wavenumber 3	1.98 / 1.57	1.64 / 1.11	1.59 / 1.05	1.90 / 1.23
Wavenumber 4	0.66 / 0.77	1.02 / 0.61	0.92 / 0.74	0.47 / 0.69

The amplitudes of wavenumber 0-4 of wind speed at 10 m for 3-km and 1-km runs averaged in the mature stage were provided in Table 7.3. In NOCPs, BMJs, and KFEXs, as the grid spacing is decreased, the amplitudes of asymmetric waves were reduced somewhat, which is consistent with the findings of Fierro *et al.* [2009]. Moreover, as suggested by Peng *et al.* [1999] and Yang *et al.* [2007], the presence of asymmetric structure is a dynamical limiting factor to TC intensity. Thereby, conversely, the weakening of asymmetric waves in the three 1-km runs could increase the storm intensity and thus the difference in TC intensity between those 3-km runs and 1-km runs. However, in GR3Ds, in contrast to the previous three CPS group cases, the amplitude of wavenumber 1 increased drastically as the grid spacing decreased from 3 km to 1 km. This may be a limiting factor to storm intensity in GR3D1, and thus a possible reason for the relatively smaller difference in TC intensity between GR3D3 and GR3D1. In addition, the larger-amplitude eyewall asymmetries in GR3D1 also explain why axisymmetric means (i.e. symmetric flow) are generally smaller in GR3D1 than those in GR3D3 (Table 7.3).

Due to the specificity of asymmetric structure in GR3Ds, it is necessary to further investigate the spatial distribution of the asymmetric structure of wind speed at 10 m in GR3Ds, which is shown in Fig. 7.8. Compared with GR3D3, although the distribution of wind speed in GR3D1 is somewhat similar to that in GR3D3, the maximum wind zone in GR3D1 is notably closer to the TC eye (Figs. 7.8a and 7.8e). The calm zone of the symmetric flow in GR3D1 is notably smaller, and the amplitude of symmetric flow in GR3D1 is significantly smaller though its maximum wind speed is larger than that in GR3D3 (Figs. 7.8b and 7.8f). The wind speed perturbations of wavenumber 1 in the two runs all have two wave peaks, namely, inner wave peak and outer wave peak (Figs. 7.8c and 7.8g). Nevertheless, the amplitude, phase, and radial position of the two wave peaks in GR3D1, especially the inner wave peak, are notably different from those in GR3D3, which are the most important reason for the difference in asymmetric structures between GR3D3 and GR3D1. The amplitude and phase of wavenumber 2 in GR3D1 are basically consistent with those in GR3D3, but similar to the situation in wavenumber 1, the radial position of wave peak in GR3D1 is notably closer to the storm center (Figs. 7.8d and 7.8h). Furthermore, the inward shifts of wave peaks in wavenumber 0-2 could also be observed in the other CPS groups of experiments, which could result in the contraction of the eyewall and favor the development of the storm.

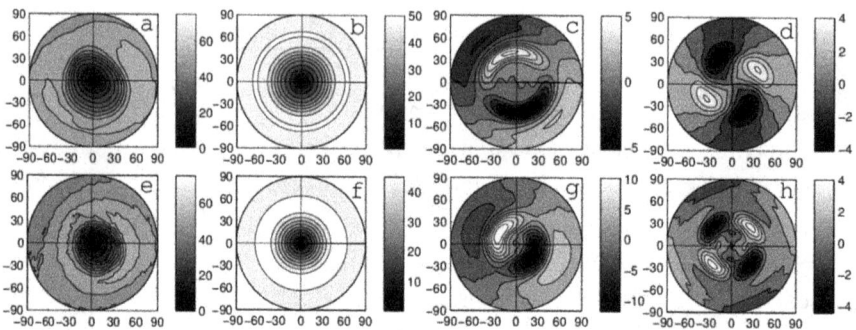

Fig. 7.8. Simulated wind speed at 10 m and its decomposition (m s^{-1}) averaged in the TC mature stage for GR3D3 (a-d) and GR3D1 (e-h). (a, e: total wind velocity; b, f: symmetric flow; c, g: wavenumber 1; d, h: wavenumber 2.). Both the abscissa and ordinate represent the distance (km) away from the TC center.

7.4.2.2 Thermodynamic structure

Latent heating is the direct force that drives the eyewall convection and the secondary circulation of a hurricane, which causes the storm to maintain itself and develop further. In this chapter, the latent heating was found to be sensitive to the choice of CPS and horizontal resolution. Fig. 7.9 shows the azimuthal- and time-averaged cross sections of the model-simulated latent heating in the TC mature stage. As expected, due to the more resolved convection, the higher-resolution runs produced larger magnitude of latent heating near the eyewall. However, compared with those in coarser-resolution runs, the simulated latent heating in higher-resolution runs were concentrated in much smaller areas closer to the eye. Hack and Schubert [1986] showed that the smaller radius at which latent heating occurs, the more significant contribution to the central pressure fall is. Thereby, the simulated latent heating at smaller radius in higher-resolution runs contributes to more central pressure falls, leading to a larger increase in the TC intensity (Figs. 7.2 and 7.3).

On the other hand, the simulated latent heating is also quite different under different CPS conditions. The four CPS groups with the most simulated latent heating, in order, are KFEXs, NOCPs, BMJs and GR3Ds, which is consistent with their rank in order of the simulated TC intensity (Figs. 7.2 and 7.3). As the grid spacing decreased from 7.5 km to 5 km, except for the BMJ cases, the simulated latent heating increased significantly, resulting in the increase of TC intensity. However, in BMJ runs, compared with BMJ7.5, the magnitude of latent heating in BMJ5 was similar to that in BMJ7.5, and the factor promoting storm intensity in BMJ5 (i.e., wider extent of latent heating) was offset by the factor limiting storm intensity (i.e., larger radius at which latent heating occurs). These may be responsible for the similarity of TC intensity between BMJ7.5 and BMJ5. Moreover, except for BMJs, there are no significant differences in latent heating and thus TC intensity between 5-km runs and 3-km runs in the other three CPS groups. The radius at which the latent heating occurs in BMJ3 is much smaller than that in BMJ5, which may contribute to the significant difference in TC intensity between BMJ5 and BMJ3. Furthermore, as the grid spacing further decreased from 3 km to 1 km, the larger magnitude and smaller radius of latent heating in 1-km runs may be responsible for stronger TC intensity in 1-km runs than in 3-km runs. Moreover, Fig. 7.9 clearly shows that GR3Ds produced the least change in the vertical distribution of latent heating especially as resolution shifted from 7.5 km to 3 km, demonstrating the least resolution-dependence of the GR3D CPS. The same result can also be found in the vertical distributions of vertical velocity and hydrometeor mixing ratio (see Figs. 7.12 and 7.13 in section 7.4.2.3).

It should be noted that, the latent heating could also affect the pressure gradient force near the eyewall and thus RMW and TC intensity, which will be discussed next.

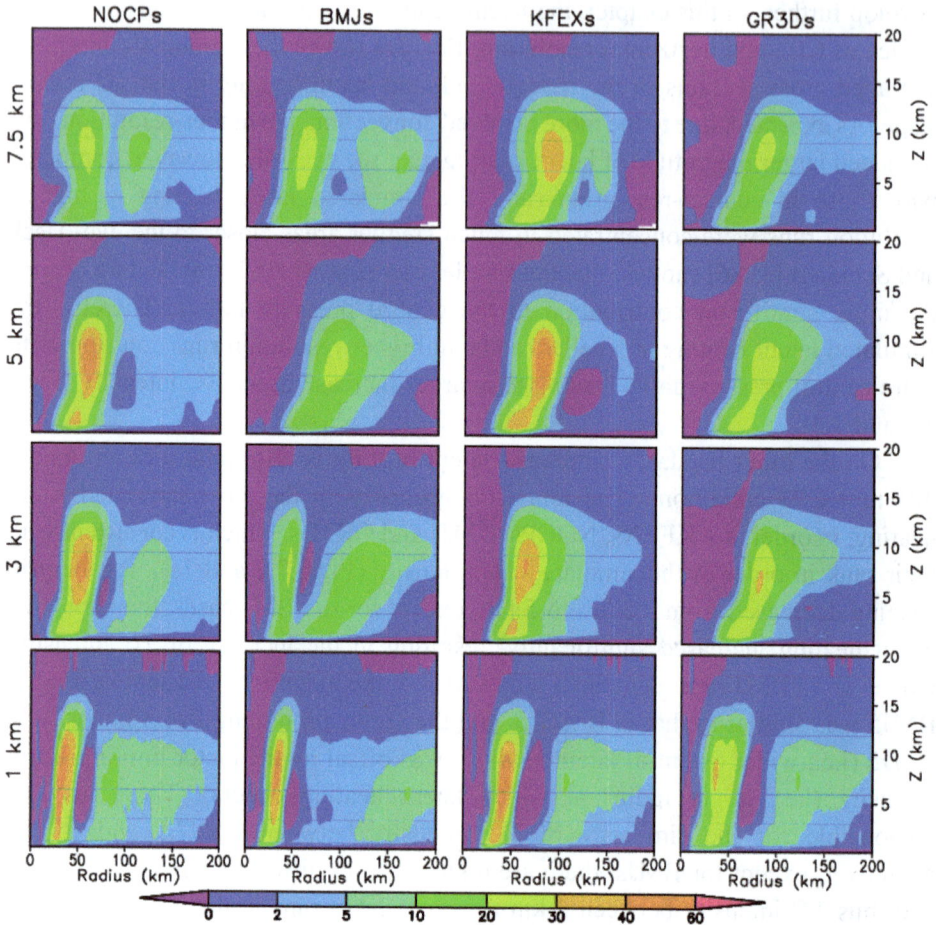

Fig. 7.9. Azimuthal- and time-averaged cross sections of the model-simulated latent heating (°C h^{-1}) in the TC mature stage.

7.4.2.3 Dynamic structure

In the previous section, we showed that the radii of eyewall decrease notably as the grid spacings decrease from 7.5 km to 1 km (especially from 3 km to 1 km). Stern and Nolan [2009] found a linear relationship between the size of the radius of maximum wind (RMW) (which basically corresponds to the radius of eyewall) and outward slope of eyewall based on both observational data analysis and theoretical deduction [Emanuel, 1986]. Moreover, many studies have further investigated the relationship between the outward slope of eyewall and TC intensity, but the results were various. Yang *et al.* [2007] showed that storms with larger eyewall tilt (or slope) were more intense because this tilt allowed more low-θ_e downdraft air to reach the subcloud inflow layer which in turn increased the air-sea entropy difference there and therefore, the energy input from the sea. They suggested that storms exhibiting less vertical tilt (smaller eyewall slope) were less intense because of enhanced inward potential vorticity (PV) mixing from the eye to the eyewall. However, Fierro *et al.* [2009] found that, despite larger eyewall slopes, the coarser-resolution runs still did exhibit similar TC intensity to those at finer resolution. Furthermore, in the work of GL10, it is implied that the TCs with a smaller eyewall slope in finer-resolution cases were notably stronger than that with a larger eyewall slope in coarser-resolution cases. Our results are basically consistent with GL10: due to the larger radii of the eyewall and thus the larger eyewall slope, the coarser-resolution cases do exhibit weaker intensity than those at finer resolution. This may be an important reason for the larger differences in TC intensity between 3-km runs and 1-km runs (Figs. 7.2 and 7.3).

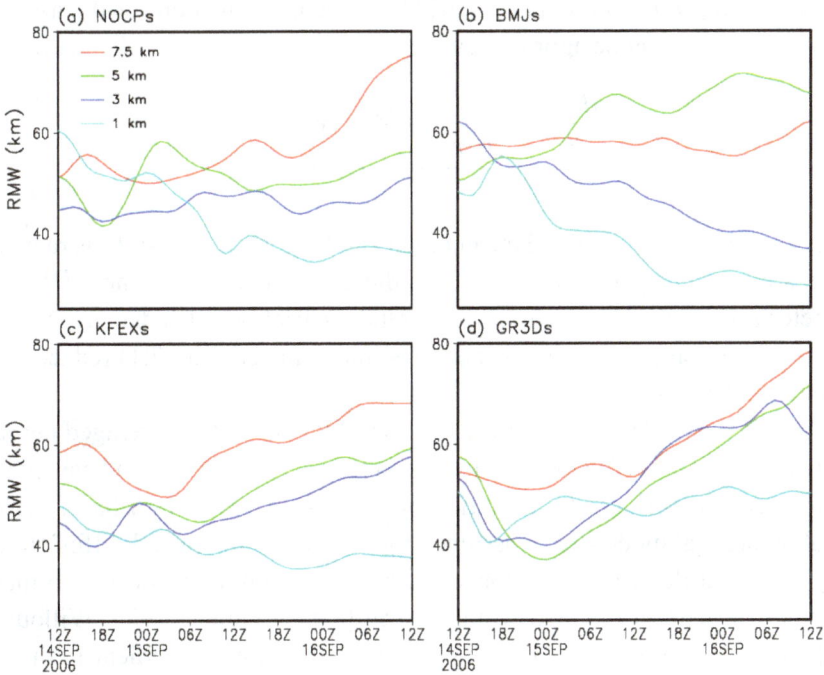

Fig. 7.10. Temporal evolutions of azimuthal mean radius of maximum winds (RMW) at 10 m for the sensitivity experiments (a) NOCP, (b) BMJ, (c) KFEX, and (d) GR3D.

To provide a more general description of the radii of eyewall, the evolution of the overall inner-core size of the simulated storms were shown in Fig. 7.10 in terms of azimuthal mean RMW at 10 m. Since the eyewall of the simulated storm can be asymmetric from time to time [Wang, 2007], we only use the azimuthal mean as a proxy of the location of the overall eyewall. Basically, it is found that the RMW increased with time in all the coarser-resolution runs (i.e., 7.5-km and 5-km runs) and some higher-resolution runs, while decreased throughout the simulation in the other higher-resolution runs (i.e. NOCP1, BMJ1, BMJ3, and KFEX1). Most of all, except for BMJ5 and GR3D3, the RMWs for the remained cases decreased notably in general as the grid spacings decreased, which is consistent with variations of model-simulated rainfall shown in Fig. 7.5 and has been noted by many studies [Yau *et al.*, 2004; Davis *et al.*, 2008; GL10].

We can see from Fig. 7.7 that, as a result of more grid-resolved convection in finer-resolution runs, the low-level pressure gradient near and outside the RMW was larger than that in coarser-resolution runs. To understand how the RMW responds to changes in the radial pressure gradient associated with grid-resolved convection, we performed a radial momentum budget analysis below. Similar to the work of *Gopalakrishnan et al.* [2011], the budget equation for the azimuthal mean radial winds can be approximated by

$$\frac{du_r}{dt} = \underbrace{-\frac{1}{\rho}\frac{\partial p}{\partial r} + \frac{v_\lambda v_\lambda}{r} + fv_\lambda}_{Term\ A} + Du_r \qquad (7.1)$$

where u_r and v_λ are azimuthal mean radial and tangential winds; r is radius; ρ and p are air density and pressure; f is the Coriolis parameter; and Du_r is the parameterized subgrid-scale diffusion, including friction, of radial winds. In the absence of friction and the forces that constitute balance, Eq. (7.1) reduces to the gradient wind equation (term A).

Fig. 7.11 provides a Hovmöller diagram of the azimuthal-averaged net radial forcing term without diffusion [i.e., term A in Eq. (7.1)] at the 100-m level. Starting with the earliest work on the evolution of a balanced vortex by Eliassen [1951], several theoretical models have assumed that the acceleration and diffusion terms in Eq. (7.1) that describe the secondary circulation may be neglected so that the vortex is in a state of gradient wind balance [e.g., Emanuel, 1986; Willoughby, 2009]. Within the inflow layer in Fig. 7.11, winds were subgradient in the outer radii but became supergradient in the eyewall region where the inflow diminished in magnitude and the convective updraft erupted, which is consistent with the results of Gopalakrishnan *et al.* [2011]. As the grid spacing decreased, updrafts were more adequately resolved, especially from 3-km to 1-km cases (Fig. 7.11).

At a grid spacing of 1 km, as suggested by GL10 and Yu and Lee [2010], features within the eyewall (i.e., an ensemble of updraft cores within the eyewall) began to be somewhat resolved. Together with the increase of the latent heating in the eyewall (Fig. 7.9) the central pressure was decreased (Fig. 7.7), leading to increased pressure gradient near the eyewall. Subsequently, under the stronger pressure gradient force and thus stronger net radial force (Fig. 7.11), the eyewall in 1-km runs would contract inward, resulting in a much smaller RMW (Figs. 7.5 and 7.10). This mechanism could also explain the reduction of RMW as grid spacing decreases from 7.5 km to 5 km. However, due to the similar grid-resolved convections for the 5-km runs and 3-km runs except for BMJs (Fig. 7.11), there is no essential difference in net radial force between 5-km runs and 3-km runs, resulting in similar RMWs. It should be emphasized that the smaller RMW in finer-resolution runs also in turn contributes to the sharper radial pressure gradient. Thereby, it appears that as the grid spacing decreases, there is a feedback between the reduction of RMW and the increase of radial pressure gradient, which contributes to the further reduction of RMW. In fact, at the finer-resolution runs, the reduced RMW has also strengthened the centripetal force in term A, thus offsetting the pressure gradient force and limiting the further reduction of RMW.

To illustrate the relationship between the outward slope of the eyewall and the size of the RMW in these simulations, Fig. 7.12 shows the azimuthal- and time-averaged cross sections of positive vertical motion and tangential velocity in the mature stage of the TC. As expected, due to the decreased RMW, the higher-resolution runs are characterized by more upright (i.e., smaller slope) eyewalls in terms of vertical and tangential velocity. *Stern and Nolan* [2009] also showed that the outward slope of the eyewall with height is directly proportional to the size of the RMW. In addition, as the grid spacings decreased, the vertical velocities in those runs were not all strengthened and even weakened (e.g., from 5-km runs to 3-km runs), but the storm intensities were almost all increased. This is consistent with the findings of Bryan *et al.* [2003] in that higher resolution could eventually weaken the magnitude of vertical motions as detrimental processes, such as entrainment with grid-scale turbulence, begin to be resolved. Thereby, as the grid spacing decreased, the intensification of TC could not be purely attributed to the strength of convection.

Comparison of the slope of eyewall (Fig. 7.12) with the TC intensity (Figs. 7.2 and 7.3) reveals a close relationship between the difference in the slope of eyewall and the difference in TC intensity. In general, the simulated storm with less (more) eyewall slope in the finer- (coarser-) resolution runs are notably stronger (weaker), which is also confirmed in GL10. Moreover, in this chapter, we found that, as the reduction of eyewall slope is not uniform across similar decrease in grid spacing, the increase in TC intensity is also not uniform. Specially, as the grid spacing decreased, if the change of eyewall slope was relatively smaller, the TC intensity would change relatively little (such as in cases of NOCP5-3, BMJ7.5-5, KFEX5-3, and GR3D5-3), while in the other cases (such as in cases of NOCP7.5-5, NOCP3-1, BMJ5-3, BMJ3-1, KFEX3-1, GR3D7.5-5, and GR3D3-1), the eyewall slope changed notably and thus a notable TC intensity change occurred. Based on the above analysis, this chapter suggested that there exist an evident and robust relationship between eyewall slope and TC intensity.

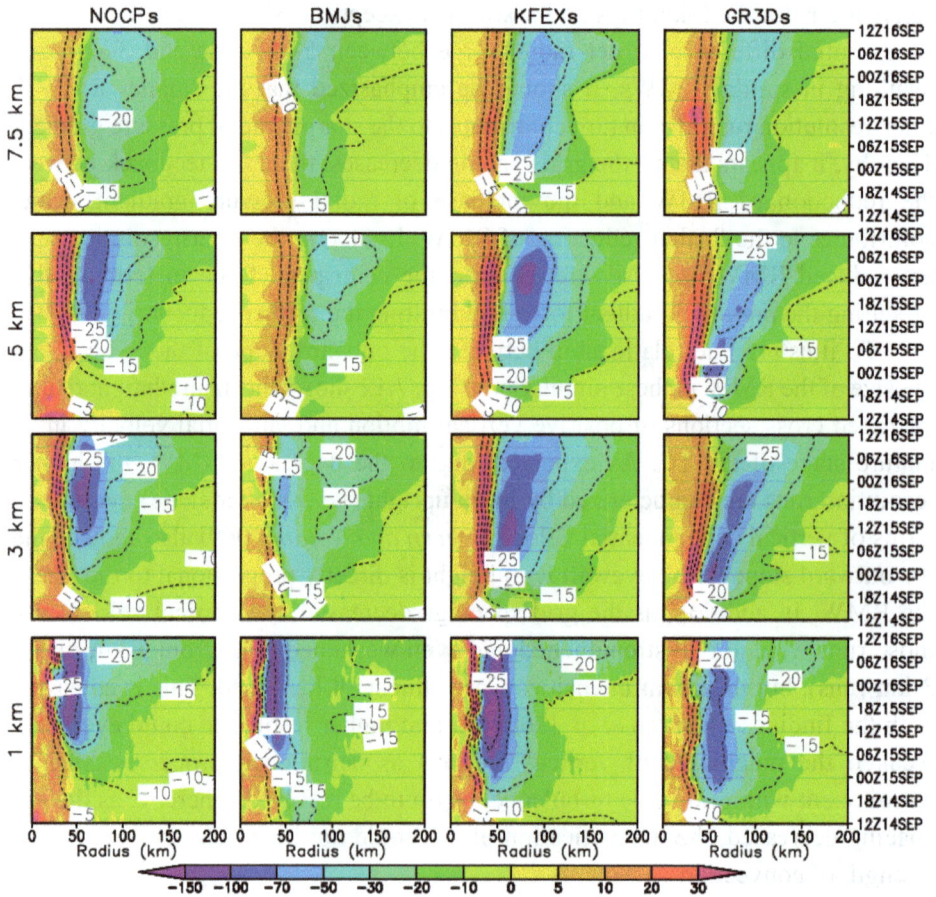

Fig. 7.11. Hovmöller diagram of the azimuthal-averaged net radial forcing term (m s^{-1} h^{-1}), excluding diffusion [term A in Eq. (7.1)] at the 100-m level. Superposed on color shades are the contours of the radial component of the velocity (m s^{-1}).

Furthermore, the relationship between eyewall slope and TC intensity may be attributed to the impact of the radial gradient. The ascending air parcels in the eyewall normally experience a reduction of the inward-directed pressure gradient force (in terms of tangential velocity), leading to an outward centrifugal displacement with increasing height, namely, the eyewall slope. In other word, the larger (smaller) eyewall slope corresponds to a less (more) upright tangential velocity contours near the eyewall (Fig. 7.12). Moreover, a less (more) upright tangential velocity often comes out with a weaker (stronger) radial gradient of tangential velocity, as suggested by Fierro *et al.* [2009], which would limit (promote) storm intensity.

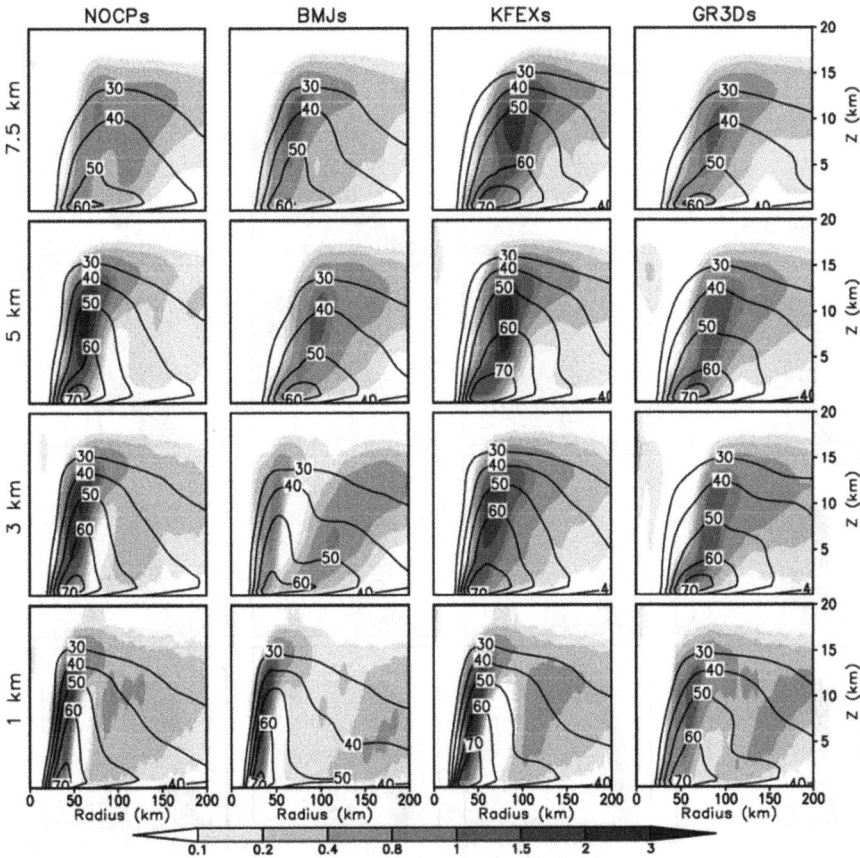

Fig. 7.12. Azimuthal- and time-averaged cross sections of positive vertical motion (m s^{-1}; shaded) and tangential velocity (m s^{-1}; contoured every 10 m s^{-1} from 30 to 70 m s^{-1}) in the TC mature stage.

7.4.2.4 *Microphysical structure*

After analyses on the asymmetric, thermodynamic and dynamic structure of the simulated storm, this section will be devoted to its microphysical structure, which will help to further identify factors promoting/limiting storm intensity and thus the reasons for the weak model convergence in simulation of TC intensity. Fig. 7.13 shows the azimuthal- and time-averaged cross section of total mixing ratio of model-simulated hydrometeors and wind vector in the mature stage of the TC. The hydrometeors are distributed mainly near the 8-km height level, which is consistent with the distribution of vertical velocity in Fig. 7.12 and the wind vectors in Fig. 7.13. Previous studies suggested that, this is because the upper-level maximum of convection could be a consequence of water unloading and buoyancy [Samsury and Zipser, 1995; Trier et al., 1996].

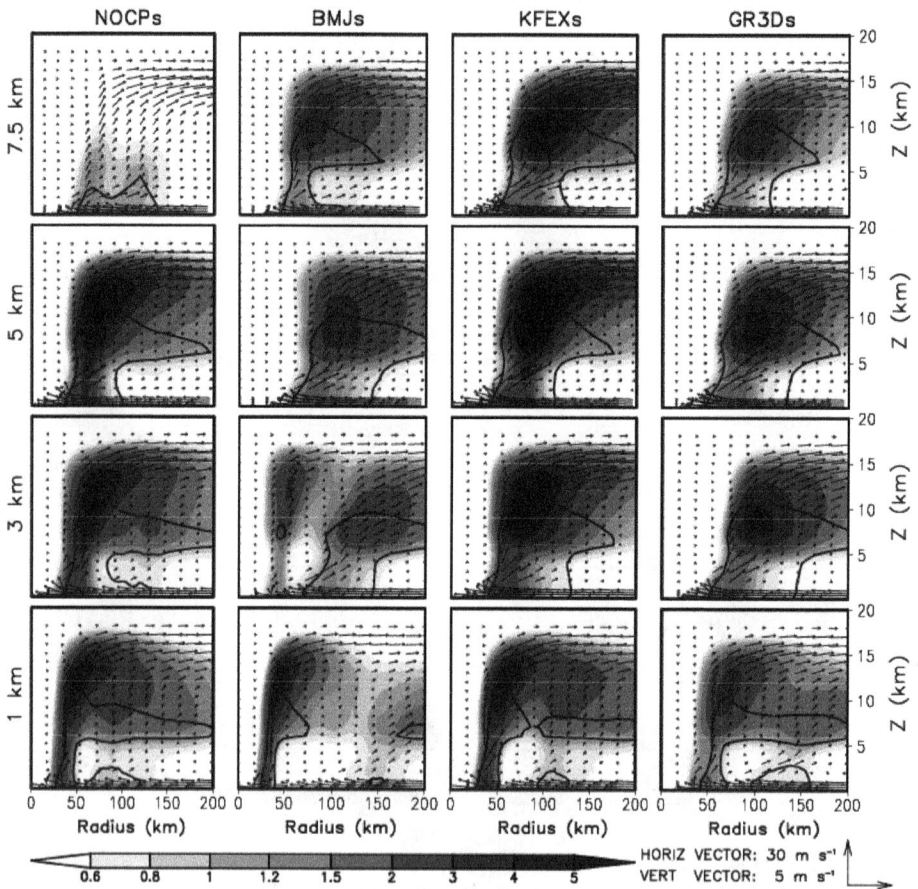

Fig. 7.13. Azimuthal- and time-averaged cross sections of total mixing ratio of model-simulated hydrometeors (g kg^{-1}; shaded) and wind vector in the TC mature stage. The thick dashed line represents the 30-dBZ contour line of simulated radar reflectivity. The wind vector is composed of radial wind and vertical velocity. Length of wind vector scale represents 30 m s^{-1} for horizontal velocity and 5 m s^{-1} for vertical velocity, respectively.

On the other hand, as grid spacing is decreased, the horizontal gradient of mixing ratio of hydrometeors and radar reflectivity became stronger and more concentrated near the eyewall, and the area occupied by the primary circulation becomes more compact in the horizontal and elongated in the vertical. Moreover, these changes were not uniform across similar decrease in grid spacing (Fig. 7.13), which is consistent with the results in Fig. 7.12. The band in NOCP7.5 showed a weak convective system with anvil clouds at a maximum height below 7 km, which is due to less grid-resolved convection in the 7.5-km runs without any CP. However, in other three 7.5-km runs (i.e., BMJ7.5, KFEX7.5, and GR3D7.5), the convective systems were fully developed with anvil clouds reaching more than 15 km in height. As the grid spacing decreased from 7.5 km to 5 km, except for KFEX5, due to enhanced grid-resolved convection, the extent and strength of the convective band were all increased markedly, which corresponded to the significant increase of TC intensity, especially in NOCP5. This result is expected, because the coarser grid spacing ($\geq$ 5 km) cannot resolve most individual cells without CP. As seen in the 3-km runs, with an exception of BMJ3, the distribution and intensity of the convective bands were basically similar to those in its 5-km runs, and thus associated with similar TC intensity.

As the grid spacing was further decreased to 1 km, the strong bands were concentrated in a narrow area near the eyewall, showing strong radial gradients and thus a stronger storm. This may be related to the weak model convergence in simulation of storm intensity at 3-km and 1-km resolutions. Furthermore, compared with the other three CPS groups, GR3D1 showed a relatively small difference in microphysical structure form GR3D3, which contributes to the similar intensity between GR3D3 and GR3D1 and thus a relatively strong model convergence.

7.4.3 *Mechanisms for the model convergence on resolution*

The possible physical mechanisms responsible for the dependence of model convergence on resolution in simulation of TC intensity are identified through comprehensive diagnostics and are schematically summarized in Fig. 7.14.

As mentioned above, it is not the extra area-integrated mass or SEF input into TC that responsible for the weak model convergence in TC intensity as the grid spacing decreased, but the change of TC structure or distribution of variables in TC system. As the grid spacing is decreased, the weaker asymmetric structure, the smaller SEF outside the eyewall, the smaller RMW, and more latent heating near the eyewall are all responsible for larger TC intensity increase, and thus weak model convergence in TC intensity (Fig. 7.14a). Furthermore, due to the smaller

RMW, the finer-resolution runs are characterized by smaller slope eyewall. Combined with the smaller SEF outside the eyewall, the more upright eyewall promotes the radial gradients near the eyewall in terms of horizontal and vertical velocity, SEF, latent heating, SLP, radar reflectivity, etc (see Fig. 7.14a). The larger radial gradients near the eyewall caused by the smaller RMW may be the main reason for the weak model convergence in TC intensity.

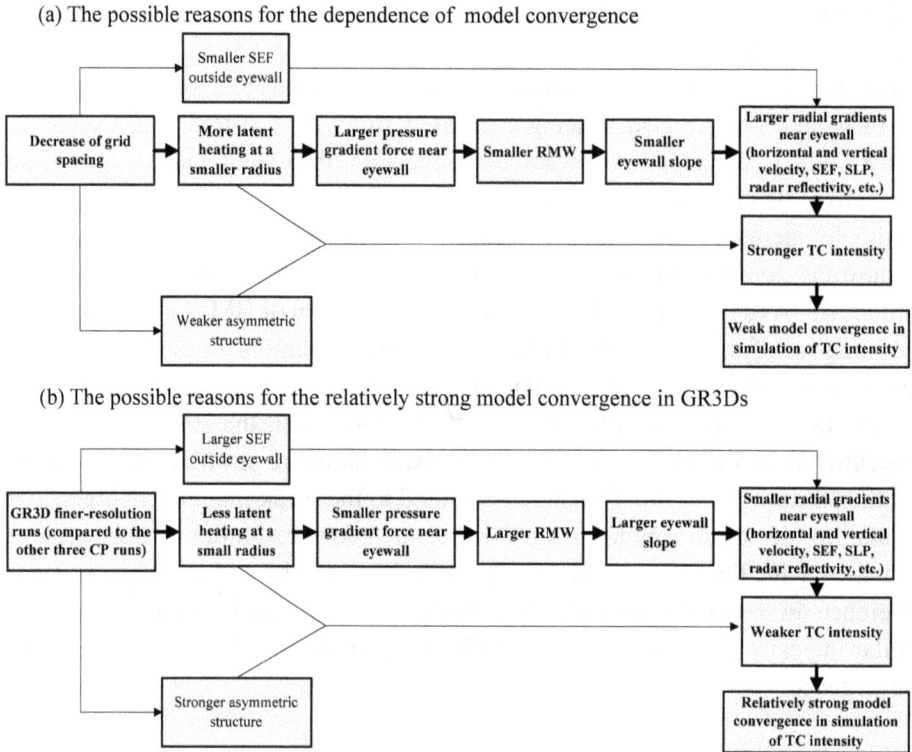

(a) The possible reasons for the dependence of model convergence

(b) The possible reasons for the relatively strong model convergence in GR3Ds

Fig. 7.14. Schematic diagram summarizing the possible reasons for the dependence of model convergence in simulation of TC intensity and the possible reasons for the relatively strong model convergence in GR3Ds.

It should be pointed out that the impacts of the factors promoting storm intensification at finer resolution are notably different under different CPS conditions. As mentioned above, the factors promoting the storm intensity in GR3D fine-resolution runs (such as GR3D1) are notably weaker than the other three CPS groups. More importantly, latent heating released from different CPS schemes exhibited different degrees of dependence on the model resolution, among which the GR3D scheme presents the least dependence while BMJ shows the most, particularly from 7.5 km to 3 km. In sum, relatively strong asymmetric structure, larger RMW and SEF outside the eyewall and less dependence of latent heat release on resolution are responsible for relatively weaker TC intensity in GR3D finer-resolution runs (Fig. 7.14b), and thus the relatively stronger model convergence in simulation of TC intensity.

7.5 Discussions

It is found that as the grid spacing is decreased, except for the notable difference between 7.5-km runs and 5-km runs, the simulated tracks are found to be little changed. Consistent with recent findings from cloud-resolving model simulations, the changes in storm intensity and structure are not uniform across similar reductions in grid spacing. There is a significant increase in TC intensity as the grid spacing decreases from 3 km to 1 km. This indicates that the model solution has not yet converged for these grid spacings. The simulations using the GR3D convective parameterization scheme present a relatively better model convergence, compared to the other three CPSs.

This is a case study. It is important to point out that, in this chapter, it is not necessary to precisely reproduce Shanshan (2006) in terms of its intensity (e.g., mean sea level pressure and maximum wind speed) and structure (e.g., spatial distribution of rainfall). Our conclusions are based on the comparisons between sensitivity experiments only, and less dependent on absolute and precise simulation results. Most importantly, in all our cases, the simulated TCs follow a similar track. This allows us to examine the sensitivity of simulated TC kinematic and microphysical structures to horizontal grid spacing and convective parameterization schemes in an environment similar to that of Typhoon Shanshan.

In this chapter, the reasons for the weak model convergence in simulation of TC intensity were discussed from perspectives of energy exchange at air-sea interface, mass flux in the boundary layer, eyewall asymmetries, and thermodynamic, dynamic and microphysical structures. Across the rather narrow spectrum of resolutions compared in detail here, there were a number of factors that limited or promoted storm intensification and thus the model convergence, as resolution varies. Factors promoting storm intensification at finer resolutions and hence weak model convergence were the smaller surface energy flux outside the TC eyewall, smaller eyewall asymmetry, more latent heating in the eyewall, and larger radial gradients in kinematic and microphysical structures, and *vice versa*. These factors are enough to overcome the negative impact of the less area-integrated SEF and mass flux on TC intensity in finer-resolution runs. Among these factors, the larger radial gradient related to the RMW and eyewall slope, is the key factor for the weak model convergence in high-resolution simulations (e.g., 1-km runs). That is, as the grid spacing decreases, due to the increased net radial force resulting from the more grid-resolved convection and enhanced latent heating near the eyewall, the RMW also decreases, leading to a more upright eyewall (smaller eyewall slope), and thus a stronger TC.

Actually, the statistical relationship between TC intensity and RMW or eyewall slope is still controversial, but for a given simulation, this relationship is robust. Based on the statistical analysis on the observation results of many storms, Stern and Nolan [2009] suggested that the outward slope of the RMW with height is directly proportional to the size of the RMW, but no relationship is found between the slope of the RMW and TC intensity. This would appear at first to contradict our results from a single TC case. In fact, for the simulation of a single storm, the slope of the RMW is linearly related to its size, as in the maximum potential intensity (MPI) theory of Emanuel [1986]; as the RMW contracts with time, the slope decreases. The intensity also appears to be very much related to slope, although the relationship is nonlinear. For a given simulation, the slope is approximately inversely proportional to intensity, which was also emphasized in Stern and Nolan [2009]. In our chapter, although the horizontal resolution and CP scheme are varied, the relationship between eyewall slope and TC intensity is still strong.

Furthermore, the roles of the factors in affecting model convergence are notably different under different CPS conditions. In contrast to those in the other three CPS groups (NOCP, BMJ, and KFEX), the TC intensity in GR3D finer-resolution runs are notably weaker than expected due to the relatively stronger asymmetric structure, larger RMW, more SEF outside the eyewall, and especially the less dependence of the GR3D scheme on model resolution. Thereby, the model convergence in GR3Ds is significantly stronger than those in the other CPS groups. The choice of CPSs indeed has a great impact on the storm structure, intensity, and thus the model convergence, so it is possible to find a reasonable CPS suitable for the higher resolution model to obtain more realistic simulations with strong convergence and thus better forecast quality.

It is interesting to make an analogy with this model convergence investigation and ensemble forecasting. In fact, this chapter tries to identify the convergent members of a group of sensitivity experiments (may be taken as an ensemble) and understand the reasons. Convergent TC intensity and structure would be a desirable attribute of perfect models, given the grid spacing and CPS changes. Convergent members may thus receive a higher weight in the ensemble. With regard to this, however, the scale issue needs to be considered. In synoptic-scale ensembles, the growth of initial errors dominates and often the consensus within the ensemble indicates higher predictability of the environment, i.e., a spread-skill relation may exist [Scherrer *et al.*, 2004]. However, for mesoscale ensembles such as in TC applications, model errors dominate and convection processes are a major contributor as nonlinearity in such processes leads to large sensitivities in the physics representations [Houtekamer *et al.*, 1996], and thus one should be cautious about using model convergence as a measure of skill. In TC applications, many more cases of validation are needed to see if model convergence actually indicates that some processes and associated nonlinearity during TC intensification dominate and likely to be true.

According to the performance of the four CPS experiments in this chapter, cases with GR3D present relatively better simulation results in terms of storm intensity and structure, especially in its high-resolution runs (such as GR3D1). However, the reason why only the GR3D scheme achieved a better performance in maintaining model convergence (i.e., what mechanism inside the GR3D scheme has contributed to the better simulation) still needs an exclusive investigation. In addition, as a new CPS designed for high-resolution simulation, GR3D has some other problems that need to be solved. We emphasize here that this chapter has analyzed the model convergence problem when some CPSs are applied to the grey-zone resolutions, however, the converged solution does not necessarily approach the truth.

Chapter 8

Mechanism of Cumulus Parameterization Scheme on Model Convergence

8.1 Introduction

The use of conventional CPS to alleviate weak model convergence is neither appropriate nor effective for models operating at high resolutions. This is because all CPSs assume that the scale of convective updraft area is much smaller than the model grid spacing, but this assumption becomes invalid when the grid spacing is a few kilometers or smaller [Grell *et al.*, 2013]. In addition, as shown in Sun *et al.* [2013a], although simulations using Grell 3D ensemble CPS [Skamarock *et al.*, 2008] exhibit a relatively stronger model convergence compared to that using other conventional CPSs, the simulated peak TC intensity at 1-km resolution is still much stronger than that at 3-km resolution. Therefore, it is necessary for further refinement of CPS in simulating TC intensity at such high resolutions.

To address the scale separation issues in convective parameterization, approaches in several CPSs have been discussed [e.g., Randall *et al.*, 2003; Gerard, 2007; Arakawa *et al.*, 2011; Grell and Freitas, 2013]. The Grell and Freitas [2013] scheme (hereafter, GF13) becomes available in the WRF model since June 2013. This scheme is based on a stochastic parameterization initially developed by Grell and Devenyi [2002]. The method proposed by Arakawa *et al.* [2011] for a smooth transition across scales is used in the GF13 scheme, which is designed for the grid spacing equal to or smaller than a few kilometers. The GF13 scheme has a capability to sufficiently remove moist instability for the entire grid point. The fractional area covered by convective updrafts (σ) in the grid cell is a key parameter in GF13. The conventional cumulus parameterizations assume $\sigma \sim 1$, at least implicitly, and define the thermal structure of the cloud environment using the grid point values of the thermodynamic variables. The proposed framework in GF13, termed as "unified parameterization", eliminates this assumption from the beginning, and allows a smooth transition to an explicit simulation of cloud-scale processes as the resolution increases [Arakawa and Wu, 2013]. It behaves similarly to conventional schemes when the updraft area is much smaller than the grid size.

As the updraft area in a grid box approaches the grid size, the parameterized subgrid convection gradually decreases [Grell *et al.*, 2013].

In this chapter, we evaluate the performance of GF13 on the aforementioned model convergence at the grey-zone resolutions based on a case study of TC Shanshan (2006) and intend to disclose the influence mechanism of CPS on the model convergence in simulations of TC at grey-zone resolution.

8.2 Experimental design

WRFV3.5 is used to simulate Shanshan (2006). The initial and boundary conditions are derived from the NCEP final analysis data (FNL). A vortex-following nested-grid configuration of WRF with three two-way nested domains (DM1, DM2, and DM3) is used in this chapter. Two extra experiments with resolutions of 1.66-km and 1-km in the inner most domain (D04) are also performed. The inner domains (DM2, DM3, and DM4) move automatically with the storm during the model integration [Skamarock *et al.*, 2008]. The detailed information of the domain settings is analogous to Table 7.1.

This study consists of five simulations with 7.5-km, 5-km, 3-km, 1.66-km and 1-km horizontal resolutions in the finest mesh. The GF13 is used in all the five simulations. To keep the consistency of simulations in a nested domain configuration, DM1 and DM2 cover the same areas in all runs. The size of DM4 in the 1.66-km and 1-km runs is roughly the same as that of DM3 in 7.5-km, 5-km and 3-km runs. The physical and dynamical options in this study are the same as that in Sun *et al.* [2013a] except for the CPS, making it convenient for us to compare the performance of GF13 with other conventional CPSs.

8.3 Performance of the Grell and Freitas cumulus parameterization scheme

Fig. 8.1 compares the storm tracks simulated in the sensitivity experiments overlaid with the JTWC best track. Similar to Sun *et al.* [2013a], the simulated storms in all cases follow a track similar to that of the JTWC, and thus avoid large differences in environmental fields between the experiments. This enables to examine the sensitivity of simulated TC intensity and structure to horizontal resolution in the environmental conditions similar to that of typhoon Shanshan.

Fig. 8.2 depicts the temporal evolutions of the simulated and the JTWC best track TC intensity based on MSLP and MWS at 10 m, and the simulated inner-core size (i.e., RMW). It clearly shows that the simulated TC intensity increases notably as the grid spacing decreases form 7.5 km to 5 km and from 5km to 3 km.

The TC intensity in the 5-km run is stronger than that in the 7.5-km run possibly because of the enhanced convection at finer grid scale, while the change of storm structure (e.g., smaller RMW in the 3-km run) explains why the simulated TC intensity in the 3-km run is stronger than that in the 5-km run, which are consistent with that of Sun *et al.* [2013a]. However, as the grid spacing further decreases to below 3 km, the simulated TC intensity doesn't show any notable increase. When the resolution decreases from 1.66 km to 1 km, the simulated TC intensity is even slightly decreased, implying a strong model convergence in TC intensity simulation. The convergence behavior of the model as shown in results of the sensitivity experiments in this chapter is distinct. The same model demonstrates a divergence behavior when utilizing other conventional CPSs instead of the GF13 in the sensitivity experiments, since the simulated TC intensity in these experiments increases sharply and with large variation when grid spacing decreases [*Sun et al.*, 2013a].

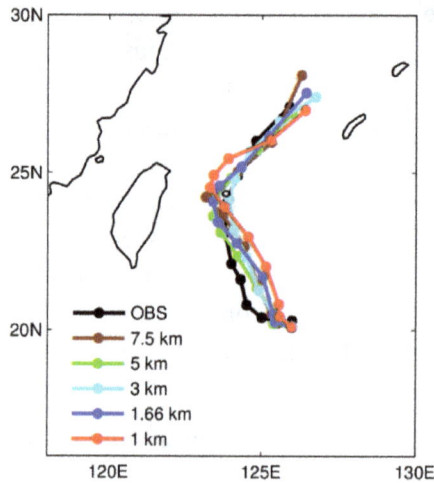

Fig. 8.1. Storm tracks overlaid with the observed best track (black disks) at 6-h intervals for the sensitivity experiments at various resolutions.

It is found that the RMW remains almost unchanged in the coarser-resolution runs (i.e., 7.5-km and 5-km runs) throughout the simulation, but it decreases during the TC developing stage (e.g., from 1200 UTC 14 September to 1800 UTC 15 September) in other higher-resolution runs (i.e., 3-km, 1.66-km and 1-km runs). Note that the decrease of RMW is consistent with the increase of TC intensity in the 3-km run compared to the 5-km run. Although statistical analyses of observations of storms found no distinct relationship between the RMW and storm intensity, the TC intensity appears to be closely related to the RMW in simulations of a single storm [Stern and Nolan, 2009; Hazelton and Hart, 2013; Sun *et al.*, 2013a; Sun *et al.*, 2014b].

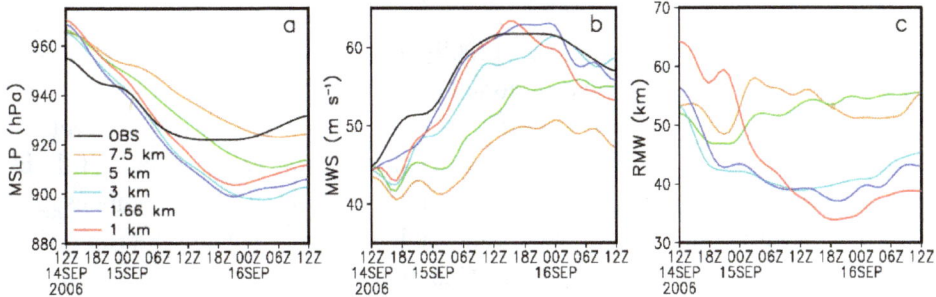

Fig. 8.2. Temporal evolutions of the simulated and the JTWC best track TC intensity based on (a) MSLP and (b) MWS at 10 m, and (c) the simulated TC inner-core size.

It should be noted that the simulated TC intensity at higher resolution is stronger than observation in terms of MSLP (Fig. 8.2a). Besides the errors in the simulations, the errors in observation may also be a reason for the bias in MSLP between the simulations and observation. On the other hand, our concern is on the model convergence, because the model convergence is a prerequisite for successful TC simulation at high resolution. Our analysis is based on the comparisons between sensitivity experiments, and less dependent on absolute and precise simulation results. The model convergence in the TC intensity simulation does not necessarily mean that the simulated TC intensity must converge to the observation. In fact, the simulated TC intensity is sensitive not only to CPSs, but also to many other model physics schemes (e.g., microphysics, boundary layer parameterization schemes, etc.). Thus, the converged finite limit could be changed due to various choices of other model physics schemes. Similar to the aforementioned conventional CPSs, not all these other schemes are appropriate for higher resolution runs. Thereby, to make the simulated TC intensity closer to the observation at high resolutions, these other schemes (e.g., microphysics, boundary layer parameterization schemes, etc.) must be improved, too. More importantly, under the condition of weak model convergence, the simulated TC intensity may be consistent with the observation at certain grid spacing, but it could also deviate from the reality at smaller grid spacing due to a larger increase in the simulated TC intensity as the grid spacing decreases. Therefore, the model with a strong convergence behavior is a prerequisite for successful TC simulation at high resolution, and the inconsistency between the simulations and observation do not affect our study of model convergence.

8.4 Possible reasons for the model convergence with a new CPS

8.4.1 *Direct impact of the CPS from the subgrid scale effects*

Wyngaard [2004] and Rotunno *et al.* [2009] suggested that, while the numerical models with a higher resolution (horizontal grid spacing ~ 1 km) can capture many of the important TC structural features (eyewall, rainbands, etc.), the effects of small-scale [~ O (100 m)], three-dimensional turbulence must still be parameterized. The drying and heating caused by small-scale compensating subsidence within one grid box may inhibit the explicit microphysical parameterizations [Grell *et al.*, 2013]. This could be a limitation factor for the storm intensification. According to the results of Rotunno *et al.* [2009], the simulated TC intensity can keep increasing with the grid-spacing of 62 m when not considering the effects of small-scale turbulence and the related convective drying and heating, which is confirmed by the results of the 1-km simulation without CPS in Sun *et al.* [2013a]. Apparently, the grid spacing of 1 km is still not fine enough to resolve the small-scale convection, and thus the subgrid convection is described by the GF13 scheme in the 1-km run.

To understand the effects of parameterized subgrid-scale subsidence in the GF13 scheme, we depict the height-time cross-sections of the heating and drying rates from the CPS averaged within a radius of 200 km of the TC center (Fig. 8.3). It can be seen that the parameterized heating and drying largely concentrate in the lower troposphere (< 5 km). This is consistent with that of Grell *et al.* [2013], which shows that large convective heating and drying rates averaged in the ITCZ over the equatorial Atlantic Ocean mainly appear near 800 hPa (about 2 km). As is shown in Fig. 8.3, the magnitude of convective heating and drying in the low levels is closely related to the TC intensity and reaches maximum in the TC mature stage (at about 2100 UTC 15 September). In order to examine the spatial distribution of convective heating and drying rates, we depict the cross sections of the heating and drying rates from the CPS in the TC mature stage in Fig. 8.4. Consistent with the results of Rotunno *et al.* [2009], the parameterized small-scale heating and drying largely distribute near the eyewall. However, the effects of heating and drying in the outer spiral rainbands (100~150 km from the TC center) cannot be ignored, especially in the simulations with smaller grid spacing. Above results clearly indicate that the small-scale turbulence in the TC region are not negligible and must be parameterized, just as suggested by Wyngaard [2004] and Rotunno *et al.* [2009].

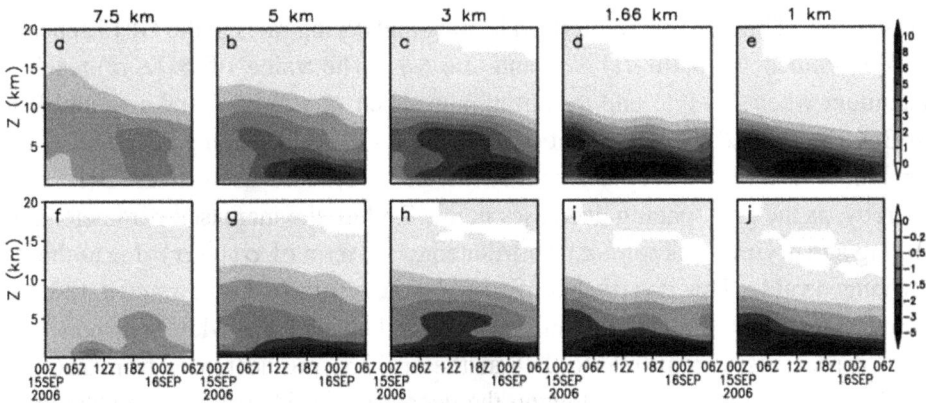

Fig. 8.3. Height-time cross-sections of the heating and drying rates from the CPS averaged within a radius of 200 km: (a-e) heating rate (°C d^{-1}); (f-j) drying rate (g kg^{-1} d^{-1}).

Here, we will give an explanation for why the simulations with GF13 scheme at higher resolution do not converge to the experiment without CPS as conducted by Sun *et al.* [2013a]. Let h be the (potential) temperature or water vapor mixing ratio, or one of their combinations such as moist static energy. Based on Grell *et al.* [2013], the scale aware eddy transport in GF13 can be given by $\overline{w'h'} = (1-\sigma)^2 (\overline{w'h'})_{adj}$, where σ is the fractional convective cloudiness and a key parameter in the GF13, $\overline{w'h'}$ is the eddy transport, and $(\overline{w'h'})_{adj}$ is the eddy transport calculated by the CPS without relaxation, namely, the solution when $\sigma \sim 1$. As suggested by Arakawa *et al.* [2011] and Grell *et al.* [2013], and σ increases as the grid spacing decreases. Moreover, according to the code of the closure of σ in GF13, σ decreases with the decrease of the grid spacing at first and then keep constant (i.e., $\sigma = 0.7$) when the grid spacing decreases below 8.47 km. As a result, as the grid spacing decreases from 7.5 km to 1 km, σ and thus $(1-\sigma)^2$ keep unchanged. This explains why the simulations with GF13 scheme don't converge to the experiments without CPS in Sun *et al.* [2013a] when the resolution is high simply because the scheme is still acting (i.e., $(1-\sigma)^2$ is finite and not zero).

The parameterized heating and drying rates increase notably as the grid spacing decreases from 7.5 km to 1 km (Figs. 8.3 and 8.4). This seems contradictory to that of Grell *et al.* [2013], namely, the parameterized subgrid convection gradually decreases as the updraft area in a grid box approaches the grid size. In fact, the great increase in heating and drying rates at 1-km run may be attributed to the inhomogeneous structure of the updraft and the environment (e.g., small-scale turbulent motion). According to the derivation of Arakawa *et al.* [2011], the eddy transport can also be expressed as $\overline{w'h'} = \sigma(1-\sigma)\Delta w \Delta h$, where Δw and Δh is the excess of w and h of the updrafts over the environment, respectively. Thus, the parameterized heating and drying rates in the GF13 scheme are determined by both $\sigma(1-\sigma)$ and $\Delta w \Delta h$. The value of $\sigma(1-\sigma)$ reaches maximum when $\sigma=0.5$, and it is minimum when $\sigma=0$ or 1 [Arakawa and Wu 2013; Grell *et al.* 2013)]. As mentioned above, as the grid spacing decreases from 7.5 km to 1 km, σ keep unchanged in GF13 scheme as well as $\sigma(1-\sigma)$. Thereby, as the grid spacing decreases below 7.5 km, the increased parameterized heating and drying rates cannot be attributed to the term of $\sigma(1-\sigma)$ due to the its unchanged value. This is further confirmed by the results of Arakawa *et al.* [2011], which shows that the correlation coefficients between the eddy transport and $\sigma(1-\sigma)$ decreases from 0.94 to 0.46 as the grid spacing decreases from 256 km to 4 km. It is obvious that, following the decrease of grid spacing, the change of the other term (i.e., $\Delta w \Delta h$) is responsible for the decreased correlation coefficients, especially at high resolution. Due to the decreased correlation coefficients between the eddy transport and $\sigma(1-\sigma)$ at high resolution, $\Delta w \Delta h$ plays a more important role in determining the parameterized heating and drying rate in terms of the eddy transport, as the grid spacing decreases from 7.5 km to 1 km. Arakawa and Wu [2013] further attributed the decreased correlation coefficients to the inhomogeneous structure of updrafts and the environment and thus the large value of $\Delta w \Delta h$. This is consistent with the results in this chapter, which shows that, the active updrafts are likely to be surrounded by weaker updrafts near the eyewall and the outer spiral rainbands at higher resolution since TC is a strong convective system with lots of small-scale [$\sim O(100m)$] turbulence, resulting in the inhomogeneous structure of updrafts and the environment. In other words, as the grid spacing decreases, the structure of the updrafts and the environment near the eyewall and the outer spiral rainbands will be more horizontally inhomogeneous, resulting a larger value of $\Delta w \Delta h$ and thus the eddy transport (e.g., the increase in parameterized heating and drying rates).

As suggested by Grell *et al.* [2013], the increase in drying and heating rates makes it harder for the microphysics to become active. The less active microphysical process in smaller grid spacing runs may limit TC intensification, resulting in the simulated TC intensity weaker than that simulated in Sun *et al.* [2013a] without CPS at the grid spacing of 1 km. Therefore, the effects of subgrid-scale drying and heating can be considered as a limiting factor for TC intensification, and thus contribute to the model convergence in the TC intensity simulations using the GF13 scheme.

Grell *et al.* [2013] proposed that the convective drying and heating caused by compensating subsidence inhibit the explicit microphysical scheme and thus can be considered as a limiting factor for TC intensification. Fig. 8.5 shows the temporal evolutions of grid scale precipitation and convective precipitation averaged within a radius of 200 km of the TC center. The temporal evolution of grid scale precipitation is similar to that of TC intensity, which keeps increasing before the TC mature stage and then decreases after the mature stage. Note that the grid scale precipitation does not increase and even decreases as the grid spacing decreases (Fig. 8.5a), which seems contradictory to the general understanding that experiments with smaller grid spacing can generate more grid scale precipitation than that with larger grid spacing. In fact, the area-averaged grid scale precipitation is determined not only by the magnitude of convection, but also by the area of precipitation. The latter is closely related to the TC inner-core size (i.e., the RMW). Comparing Figs. 8.5a and 8.2c, we can find a high correlation between grid scale precipitation and the RMW averaged in the TC mature stage with the correlation coefficient of 0.89. Therefore, the decrease of grid scale microphysical precipitation in high resolution run can be attributed not only to the increase of convective drying and heating, but also to the decrease of the TC inner-core size.

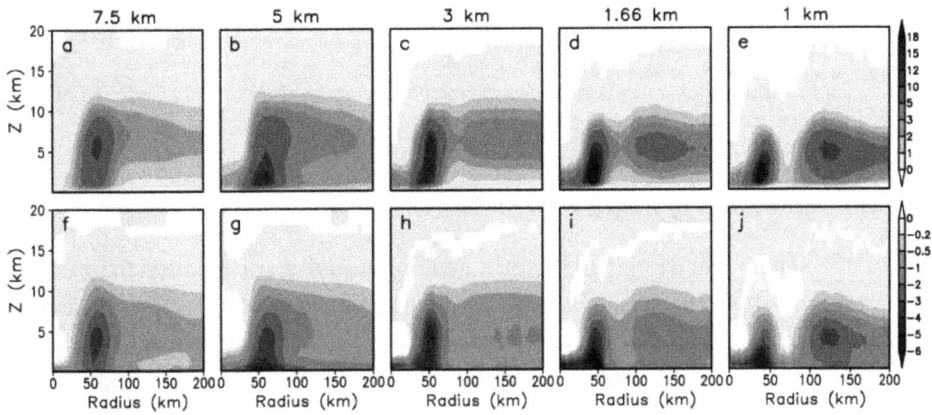

Fig. 8.4. Azimuthal- and time-averaged cross sections of the heating and drying rates from the CP in the TC mature stage: (a-e) heating rate (°C d^{-1}); (f-j) drying rate (g kg^{-1} d^{-1}).

In contrast to grid scale precipitation, the cumulus precipitation increases as grid spacing decreases (Fig. 8.5b), although the decrease of the TC inner-core size and thus the area of cumulus precipitation actually reduce the area-averaged cumulus precipitation. This is somewhat different from the results of Grell *et al.* [2013], which suggest that the effects of CPS become small as the grid spacing decreases from 20 km to 5 km. However, TC is a strong convective system with lots of small-scale [~O (100 m)] turbulence that are substantially different from the convective systems studied in Grell *et al.* [2013]. Rotunno *et al.* [2009] have revealed that the simulated flow structure of TC at a grid spacing of 62 m is characterized by vigorous, small-scale eddies within the annulus of strong winds.

In our present chapter, the effects of parameterized small-scale eddies in the GF13 scheme are proportional to the aforementioned $\Delta w \Delta h$, which increases as the grid spacing decreases due to the inhomogeneous structure of updrafts and the environment. As a result, the effects of the CPS (i.e., GF13) increase rather than decrease as the grid spacing decreases, which further demonstrates the importance of small-scale turbulence in simulating TC intensity at high resolution.

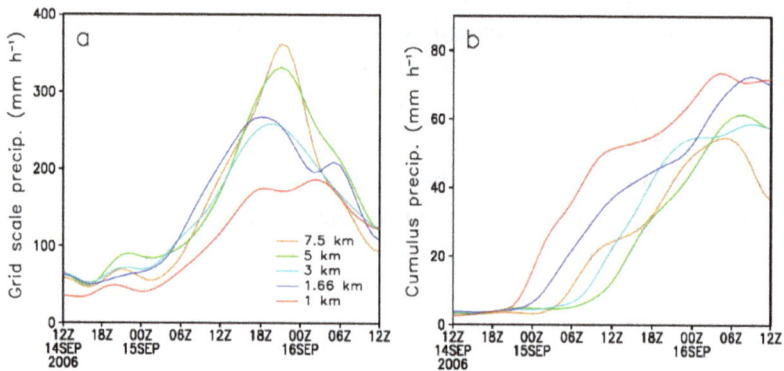

Fig. 8.5. Temporal evolutions of (a) grid scale precipitation rate and (b) cumulus precipitation rate averaged within a radius of 200 km.

8.4.2 *Indirect impact of the CPS from the TC structure*

The GF13 scheme is designed for grid spacing equal to or smaller than a few kilometers. Its capability to sufficiently remove moist instability for the entire grid point is substantially different from other conventional CPSs. Convective available potential energy (CAPE) currently serves as one of the standard measures of instability [Moncrieff and Miller 1976]. Fig. 8.6 shows the radial distribution of CAPE averaged in the mature stage of TC. Due to the latent heating and modulation of CP, CAPE is effectively removed over areas close to the TC eyewall. Such an effective CAPE removal suggests a fundamental difference between convection over the TC eyewall area and that within the rainbands. As the grid spacing decreases from 7.5 km to 3 km, CAPE also decreases notably near the eyewall but increases outside. This feature is especially distinct when the grid spacing decreases from 5-km to 3-km. Since the GF13 scheme can effectively remove moist instability near the eyewall in experiments with small grid spacing, it eventually prevents the storm from overly intensifying in the high-resolution runs. However, as the grid spacing further decreases from 3 km to 1 km, there is no significant change in the radial distribution of CAPE except for a notable decrease outside the TC eyewall in the 1-km run (Fig. 8.6). Above result indicates that the CAPE-removal ability of the GF13 scheme converges to a finite limit as grid spacing decreases, which is consistent with the results of Grell *et al.* [2013].

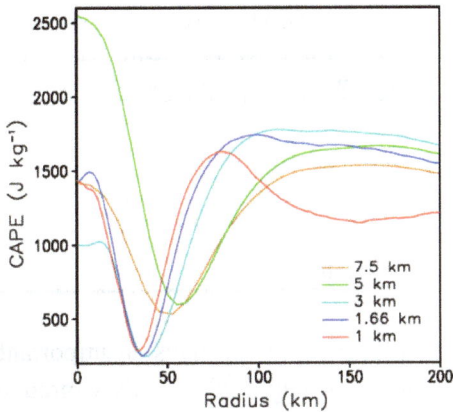

Fig. 8.6. Radial distributions of the azimuthal-averaged CAPE averaged in the TC mature stage from 1800 UTC 15 September to 0600 UTC 16 September.

It has been pointed out that the system-scale storm development is a large-scale response to the net vertical mass flux driven by diabatic heating in the convective cores [Hendricks *et al.*, 2004; Montgomery *et al.*, 2006], which leads to the enhancement of the system-scale secondary circulation in a process akin to the Eliassen balanced vortex model forced by heat sources. The spatial-averaged inflow and vertical mass fluxes, and integrated boundary layer (BL) kinetic energy in the experiments using the GF13 at different horizontal resolutions are listed in Table 8.1. It can be seen that, as the grid spacing decreases from 7.5 km to 5 km, the simulated storm exhibits a larger inflow mass flux in the low levels due to more grid-resolved convection. The larger inflow mass flux, by virtue of mass conservation law, is consistent with larger vertical mass flux and contributes to a larger BL kinetic energy. However, as the grid spacing decreases to smaller values, the secondary circulation becomes weaker at higher resolution (especially from 5-km run to 3-km run) in terms of averaged inflow and vertical mass fluxes and BL kinetic energy. A weaker secondary circulation is unfavorable for storm intensification, possibly due to the smaller inner-core size and thus the smaller size of strong wind region in higher-resolution runs. Yet the simulated storm intensity increases as the grid spacing decreases from 5 km to 3 km, indicating that these conditions unfavorable for storm intensification (e.g., the weaker secondary circulation) must be offset by other conditions favorable for storm development (e.g., the smaller radius at which strong latent heating occurs and thus larger radial pressure gradient). We will discuss this issue later.

Table 8.1. Time-averaged inflow and vertical mass fluxes (10^9 kg m s^{-1}) out of a cylinder of radius of 100 km and 1-km height from TC center in the TC mature stage from 1800 UTC 15 September to 0600 UTC 16 September. The rightmost column shows the time-averaged BL integrated kinetic energy (10^{13} m^2 s^{-2}) within a cylinder of 1-km height and a radius of 200 km from TC center in the TC mature stage.

Cases	Inflow mass flux	Vertical mass flux	BL kinetic energy
7.5-km run	-6.90	7.75	3.84
5-km run	-10.85	10.61	4.88
3-km run	-7.63	7.95	4.53
1.66-km run	-5.51	6.78	4.33
1-km run	-4.20	6.66	4.39

Vertical velocity in the TC eyewall plays an important role in determining vertical mass flux and thus the strength of secondary circulation. To examine the distribution of vertical velocity in the eyewall over the entire vertical extent of the TC, hourly contoured frequencies by altitude diagrams [CFADs; Yuter and Houze, 1995] for each run are computed to produce a time-averaged CFAD in the TC mature stage (Fig. 8.7). Due to more grid-resolved convection at higher resolution, the magnitude of the strongest updraft (downdraft), indicated by the 0.1% contour, increases by approximately 3 m s^{-1} (2 m s^{-1}) as the grid spacing decreases from 7.5 km to 3 km. Also, as resolution increases, a broader distribution of vertical velocity is present in the 3-km run. However, there is no significant increase in the magnitude of strong updraft and no broadening of the range of vertical motions as the grid spacing further decreases from 3 km to 1 km. This may be a reason for the strong model convergence using the GF13. In addition, as the grid spacing decreases from 1.66 km to 1 km, the magnitude and proportion of strong updraft, indicated by the 2% contour, only reduce slightly. This fact partly explains why the TC intensity decreases in the 1-km run compared to that in the 1.66-km run.

To disclose why the model convergences in the GF13 simulations, we depict the cross sections of latent heating rate and radial pressure gradient in the TC mature stage (Fig. 8.8). It shows that the latent heating is the direct forcing that drives the eyewall convection and the secondary circulation of a TC, which subsequently enables the TC to maintain itself and further develop [Kuo, 1965]. Not unexpectedly, the model produces stronger microphysical latent heating near the eyewall due to more grid-resolved convection and thus a stronger storm as the grid spacing decreases from 7.5 km to 3 km (especially from 7.5 km to 5 km). However, the magnitudes of latent heating and pressure gradient remain unchanged as the grid spacing further decreases from 3 km to 1 km. The little changes of latent heating and pressure gradient between the 3-km run and 1-km run with the GF13 scheme may be attributed to the large amount of convective drying and heating (Figs. 8.3 and 8.4), which inhibits the explicit microphysical parameterizations [Grell *et al.*, 2013]. In contrast, both latent heating and pressure gradient increase significantly when using other conventional CPSs and when grid spacing decreases from 3 km to 1 km [Sun *et al.*, 2013a]. Such a significant increase in latent heating and pressure gradient is one important reason for the intensification of TC in the 1-km run. As suggested by Hack and Schubert [1986], a stronger latent heating within a smaller radius contributes to a greater central pressure fall. This helps explaining the notable increase in radial pressure gradient and central pressure fall as the RMW (approximately the radius at which strong latent heating occurs) decreases notably from 5-km run to 3-km run. Contrary to the results of Sun *et al.* [2013a], the RMW remains unchanged in our present chapter and the storm intensity changes little as the grid spacing decreases from 3 km to 1 km.

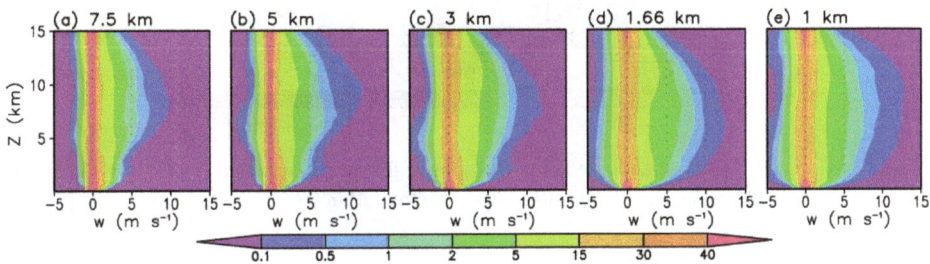

Fig. 8.7. Composite CFADs (shading, %) of the vertical velocity in the TC mature stage from 1800 UTC 15 September to 0600 UTC 16 September. CFADs are taken from points within a radius of 60 km from the TC center for 7.5- and 5-km runs, and 50 km for 3-, 1.66- and 1-km runs.

The magnitudes and distributions of vertical motion and tangential velocity are basically consistent with that of latent heating and radial pressure gradient (Fig. not shown). This is because the eyewall updraft and thus the secondary circulation is driven by the latent heating [Kuo, 1965; Willoughby, 1988], while the tangential velocity is proportional to the radial pressure gradient under the assumption that the vortex is in gradient wind balance [Willoughby, 1990]. Note that the model with grid spacing below 3 km produces a moat and an outer spiral rainbands outside the eyewall in terms of latent heating (Fig. 8.8). The moat is determined as a region of the strain-dominated flow outside the RMW, where essentially all fields are filamented and deep convection is supposed to be highly distorted and even suppressed [Rozoff *et al.*, 2006]. However, instead of suppressing deep convection, the strain flow in moat area provides a favorable environment for the development of organized inner spiral rainbands and thus contributes to the storm intensification

[Wang, 2008]. Nonetheless, as the grid spacing decreases from 1.66 km to 1 km, the radial pressure gradient in the outer spiral rainbands increases due to the larger latent heating and surface pressure decreases as a result of hydrostatic adjustment (Figs. 8.8d and 8.8e). Consistent with the results of Wang [2009], the decrease in surface pressure under the outer spiral rainbands reduces the radial pressure gradient across the RMW and the storm intensity also decreases correspondingly in terms of maximum wind at lower troposphere in 1-km run (Fig. 8.2b). Thereby, the larger latent heating in the outer spiral rainbands is one reason for the slight decrease in storm intensity as the grid spacing decreases from 1.66-km to 1-km.

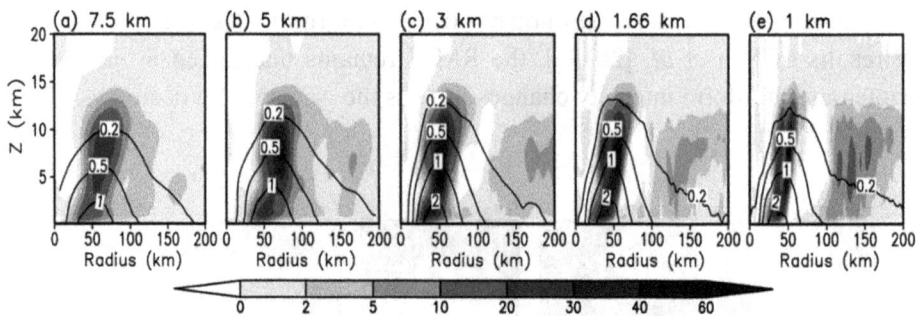

Fig. 8.8. Azimuthal- and time-averaged cross sections of the model-simulated latent heat rate (°C h^{-1}; shaded) and radial pressure gradient (10^{-2} Pa m^{-1}; contoured) in the TC mature stage.

Overall, as the model resolution varies, there are a number of factors that either limit or promote storm intensification and affect the model convergence. These factors include the convective heating and drying, the strength of secondary circulation, the magnitude and proportion of strong eyewall updrafts, the magnitude of the latent heating in eyewall, the RMW and the radius at which strong latent heating occurs, and the activities of moat and outer spiral rainbands. Compared with simulations using other conventional CPSs in Sun *et al.* [2013a], a stronger model convergence (e.g., fairly small change of TC intensity from 3-km to 1-km resolution runs) is found in present study due to less changes of these factors.

8.5 Concluding remarks

In this chapter, the WRF model is used to evaluate the performance of a new convective parameterization scheme (i.e., the GF13) on model convergence in simulations of a tropical cyclone at grey-zone resolutions. Compared with the performance of other conventional CPSs, the GF13 leads to a stronger model convergence in simulations of the TC intensity as the grid spacing decreases from 3 km to 1 km. Namely, a small difference in the simulated TC intensity occurs between 3-km and higher resolution runs using the GF13, indicating that the model performance is independent of horizontal resolutions. The model convergence is a desired attribute that would increase our confidence in the simulations. From this point of view, the GF13 performs better than other conventional CPSs in simulating the TC intensity at grey-zone resolutions, at least for the case of Shanshan (2006).

Due to the significant effects of small-scale turbulence in the TC eyewall and rainbands, the convective heating and drying described in the GF13 increase as the grid spacing decreases. This may inhibit the explicit microphysical parameterization and thus suppress the TC intensification. Different from previous studies that focus on the direct impact of convective heating and cooling, we pay more attention to their indirect impact on TC intensity in the point of view of the TC structure. As the grid spacing decreases from 7.5 km to 5 km, the increase of TC intensity can be attributed to a stronger secondary circulation, a larger magnitude and proportion of strong eyewall updraft, and a larger magnitude of the latent heating in the eyewall. As the grid spacing decreases from 5 km to 3 km, the RMW decreases and the large latent heating over the reduced RMW results in a central pressure fall, a larger radial pressure gradient and an increase in TC intensity. However, the simulated TC intensity changes slightly as the grid spacing further decreases from 3 km to 1 km because no significant changes are found in the RMW and in other terms related to the storm structure (i.e., eyewall updrafts, activities of moat and outer spiral rainbands). The slight changes in the simulated TC intensity at such high resolutions indicate a great model convergence. Therefore, the GF13 presents an appropriate option to solve the problem of weak model convergence in simulating TC intensity at grey-zone resolution.

Chapter 9

Effects of Inner and Outer Sea Surface Temperature on Tropical Cyclone Intensity

9.1 Introduction

Over the tropical oceans, low-level sensible and latent heat fluxes are important energy sources for TCs. Therefore, SST is a vital factor that determines TC genesis and intensification [Emanuel, 1986; Rotunno and Emanuel, 1987; Holland, 1997; Persing and Montgomery, 2005; Bell and Montgomery, 2008]. Placing the energetics of a TC in the framework of a Carnot cycle, Emanuel [1995] demonstrated that high SST could lead to a strong TC with high potential intensity.

   Further studies based on numerical models attempt to find a radius within which the surface enthalpy flux (the sum of sensible heat and latent heat fluxes) plays a vital role in determining the intensity of the TC [Xu and Wang, 2010b; Miyamoto and Takemi, 2010; Sun et al., 2013a]. Although the size of this radius is controversial, results of above studies have shown that the impact of the surface enthalpy flux and SST outside this radius on TC intensity is substantially different from that inside the radius. This radius is about 7-8 times the radius of the maximum wind (RMW) in the axisymmetric model of Miyamoto and Takemi [2010]. Farther outward, the surface enthalpy flux doesn't affect the TC intensity. As suggested by Xu and Wang [2010b] (hereafter XW10), this radius is about 2-2.5 times the RMW and the surface enthalpy fluxes outside this radius are crucial to the growth of the TC inner-core size but reduce the TC intensity. Similar to results of XW10, Sun et al. [2013a] demonstrated that the inner SST within a radius 2-3 times the RMW contributes greatly to the increase in TC intensity while the outer SST outside this radius reduces TC intensity. More important, this radius may be useful to estimate when a TC is influenced by a sudden change in fluxes while it is approaching a warm or cold eddy.

   Previous studies implied that the high sensitivity of TC intensity to the inner SST can be attributed to the negative effect of the outer SST on the TC intensity [Miyamoto and Takemi, 2010; XW10; Sun et al., 2013a]. These studies are more focused on the different effects of inner and outer surface enthalpy fluxes on TC intensity, but less involved in exploration of the possible mechanisms. One in-depth research topic on this issue is to Fig. out the mechanisms that favor or suppress the TC development as the inner and outer SST changes. Sun *et al.* [2013a] have noted the importance of the inner and outer SST on TC intensification based on energy transfer between the air and sea and revealed its importance in TC intensification [e.g., Emanuel, 1991b; Holland, 1997]. Unfortunately, they underestimate or ignore the role of TC eyewall structure on determining TC intensity as the SST changes, i.e., the SST can affect the TC intensity by influencing the TC eyewall structure. The former impact emphasizes the change of the surface enthalpy fluxes, while the latter impact focuses on the change in the eyewall structure. The eyewall of a TC is a ring with deep convection near the RMW. The TC obtains most of its energy from heat fluxes transferred from the ocean surface to the atmosphere, which is released later as latent heat due to moist convection in the eyewall [Wang and Wu, 2004]. The eyewall structure is closely related to the TC intensity (e.g., minimum surface level pressure and maximum wind speed) and influenced by the underlying SST [e.g., Chen *et al.*, 2010; Lee and Chen, 2012]. There is a strong dependence of the simulated storm size on the surface enthalpy fluxes outside the eyewall. The results are confirmed by further study of XW10. Nevertheless, both Wang and Xu [2010] and XW10 did not address the relation between changes in storm intensity and that in storm structure (e.g., RMW, radial pressure gradient, etc.). According to the sensitivity experiments in Fierro *et al.* [2009] and Sun *et al.* [2013b], the simulated storms with smaller RMW are often accompanied by larger radial pressure gradients near the eyewall. This larger radial pressure gradient is favorable for TC intensification. However, previous studies have not explicitly revealed whether a storm over an ocean warm pool exhibits structural features that favor TC intensification. Thereby, the impact of SST on TC intensity in view of storm structure still remains a hypothesis that needs further studies.

It should be mentioned that the reasonable design of sensitivity experiments is a prerequisite to get reliable results on the response of TC intensity to SST change. In their sensitivity experiments of Miyamoto and Takemi [2010] and XW10, the surface enthalpy fluxes outside certain radii are eliminated to determine the radius inside which the enthalpy flux is of great impact on TC intensity. However, although the underlying SST determines TC intensity by feeding the surface enthalpy flux to the TC, the change of SST may not be consistent with the change of the surface enthalpy flux induced by SST change. To better understand the TC intensity change especially when a TC is approaching an ocean warm or cold pool, it is more appropriate to conduct a series of sensitivity experiments on SST change rather than on the surface enthalpy flux change. In the ten sensitivity experiments of Sun *et al.* [2013a], the change of SST is modified with a weighting function that is maximized in the TC center. SST changes linearly outwards until the change becomes zero at a prescribed radius and remains zero further outwards. Due to the non-uniform change of SST within the certain radius, the design of their weighting function is somewhat inappropriate and inefficient to estimate the difference in SST anomalies and their impacts on TC intensity between the sensitivity experiments. In this study, we carefully design the sensitivity experiments to overcome the weaknesses of previous studies.

Although the importance of the local SST and the surface enthalpy flux to TC intensity has been addressed, questions remain as to 1) how will TC activity (e.g., the intensity, inner-core size and structure of TC) respond to the SST change over different radial extents with respect to its center, and 2) what are the possible reasons for the different impacts of SSTs over different radial extent on TC intensity from the viewpoint of the surface enthalpy flux and eyewall structure. While the emphasis of other studies [e.g., Miyamoto and Takemi, 2010; XW10; Sun *et al.*, 2013a] was mainly on the impact of SST on TC intensity in view of the surface enthalpy flux, this study give a more comprehensive view that also includes the impact of SST in view of storm structure by examining the near-eyewall structure and upper-level temperature. To investigate the impact of SST on TC activity, a suite of sensitivity experiments are designed for the case of Typhoon Shanshan (2006). The WRF model forced by artificially changed SSTs is used to simulate Shanshan.

9.2 Experiment designs

The model used in this study is the WRF-ARW. The model domain is triply-nested with two-way interactive nesting. The inner domains, which are designed to define Shanshan, automatically move to follow the position of model storm via automatic vortex-following algorithm [Skamarock *et al.*, 2008]. The initial and lateral boundary conditions are extracted from the $1° \times 1°$ NCEP final analysis data (FNL) at 6-h intervals. The TC Bogus scheme in the WRF model is adopted to generate the initial field [Skamarock *et al.*, 2008]. In this study, daily SST is updated using the TRMM Microwave Imager (TMI) level-1 standard product. It is a gridded data with $0.25° \times 0.25°$ resolution.

The model integration starts at 0000 UTC 14 September 2006 and ends at 1200 UTC on 16 September 2006, which covers the intensification and the steady-state period of Shanshan (hereafter, CTR). The Yonsei University non-local-K planetary boundary layer scheme [Hong *et al.*, 2006] and Monin-Obukhov surface layer scheme [Dyer and Hicks, 1970; Paulson, 1970; Webb, 1970; Beljaars, 1994] are used in this study. Some other important physical schemes include Lin microphysical scheme [Lin *et al.*, 1983], and Betts-Miller-Janjić convective parameterization scheme [Betts, 1986; Betts and Miller, 1986; Janjić, 1994; Janjić, 2000]. Details of the model configuration and simulation results can be found in Sun *et al.* [2012]. The triply-nested model is run with grid spacing of 45, 15, and 5km, respectively and the model output is saved at 5-min interval. Statistical analysis of the simulation results in the inner-most domain (D3) shows that the average track error is about 55 km, MSLP error is 4.7 hPa and the maximum wind speed error is 4.4 m s^{-1}. Figs. 9.1a and 9.1b show the triply-nested model domains and the tracks of Shanshan at 6-h intervals from the observation and CTR, respectively.

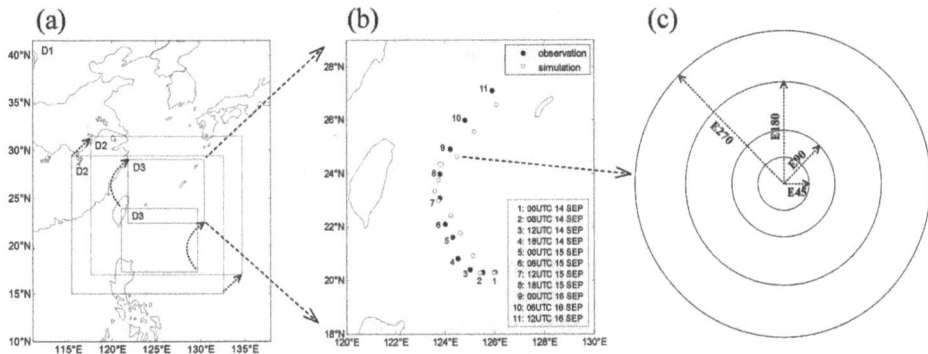

Fig. 9.1. (a) Triply-nested model domains; (b) the observed and simulated track of Shanshan (2006) at 6-h intervals (b); and (c) the schematic of extents and radii of SST change in sensitivity experiments.

In addition to the CTR, 21 sensitivity experiments are performed to investigate the response of TC intensity to the changes of SST over different radial extents. In these experiments, the underlying SST is either reduced or increased within a specified radius of the TC center in all domains (D1, D2 and D3). In particular, the SST is modified with a weighting function that is maximized within a prescribed radius of the TC center and decreases linearly outward to zero at the prescribed radius plus 15 km, and remains zero farther outward (This extra 15-km-smoothing is not necessary and not performed in D1). Note that, since the TC is a moving system, the SST field is modified on the basis of TRMM/TMI SST data in each time-step to ensure the center of the changed SST region is always consistent with the TC center in our sensitivity experiments. The extents and radii of some sensitivity experiments are shown in Fig. 9.1c.

The effects of underlying SST on intensity, inner-core size and structure of the TC are investigated through comparing results of the sensitivity experiments. Table 9.1 lists the sensitivity experiments and their simulated maximum intensities and RMWs. It should be pointed out that, such artificially moving SST anomalies in our sensitivity experiments may not exist in reality. In addition to SST, ocean heat content and mixed layer depth could also be important to storm intensity and structure due to their influence on the strength of the storm-induced SST cooling. Since it is the SST in the WRF model that directly influences the energy transfer between the ocean and atmosphere, results of our experiments with the imposed SST changes described above will help answering the question that, within what radial extent of a storm is the underlying SST relevant for determining TC intensity and how does TC intensity change as the TC approaches an ocean cold or warm pool? In addition, in all our sensitivity experiments, the simulated TCs follow a similar track, and thus prevent the differences in TC structure to be due to different environmental fields between the experiments. This allows us to examine the sensitivity of simulated TC intensity and structure to the underlying SST in an environment basically similar to that of Typhoon Shanshan.

Table 9.1. Summary of the control experiment and sensitivity experiments. Variable r is the radial distance from the storm center; ΔSST is the artificial modification of SST. MSLPmin and Vmax are the lifetime minimum MSLP and lifetime maximum wind speed at 10 m for each simulation, respectively. RMWave is the RMW at the 10 m averaged in the last 36-h simulation (from 0000 UTC on 15 September to 1200 UTC on 16 September) when the difference between sensitivity experiment and CTR is significant.

Exp.	Modification to the underlying SST	$MSLP_{min}$ (hPa)	V_{max} (m s^{-1})	RMW_{ave} (km)
CTR	Control experiment	920.6	53.3	55.0
E45-2	ΔSST = -2 for r≤30 km, and linearly increases to 0℃ at r=45 km	923.4	51.6	62.4
E90-2	ΔSST = -2 for r≤75 km, and linearly increases to 0℃ at r=90 km	939.6	48.8	82.1
E180-2	ΔSST = -2 for r≤165 km, and linearly increases to 0℃ at r=180 km	950.7	45.6	108.8
E270-2	ΔSST = -2 for r≤255 km, and linearly increases to 0℃ at r=270 km	934.0	45.7	38.7
Eall-2	ΔSST = -2℃ for all the simulation domain	941.3	45.9	45.0
E45+2	ΔSST = 2 for r≤30 km, and linearly decreases to 0℃ at r=45 km	900.3	58.9	47.8
E90+2	ΔSST = 2 for r≤75 km, and linearly decreases to 0℃ at r=90 km	889.3	61.8	47.1
E180+2	ΔSST = 2 for r≤165 km, and linearly decreases to 0℃ at r=180 km	907.9	59.2	60.8
E270+2	ΔSST = 2 for r≤255 km, and linearly decreases to 0℃ at r=270 km	915.6	54.6	83.1
Eall+2	ΔSST = 2℃ for all the simulation domain	905.6	52.8	82.8

9.3 Sensitivity to SST anomalies over different radial extent

9.3.1 *Storm intensity*

To demonstrate the sensitivity of TC intensity to SST anomalies over different radial extents, we compare in Figs. 9.2 and 9.3 the temporal evolutions of MSLP and maximum surface wind speed simulated by the sensitivity experiments. Note that the evolutions of TC intensity for the sensitivity experiments with 1°C SST change are not shown because the change of TC intensity is gradual as the magnitude of SST change varies from 1°C to 2°C. Since the RMW is about 55 km in CTR (Table 9.1), E45 can be considered as a sensitivity experiment for the SST change in the TC eye region. In contrast, in E90 the SST change occurs in both the eye and eyewall regions. E180 and E270 can be regarded as the experiments with SST changes in the inner and outer spiral rainbands, respectively.

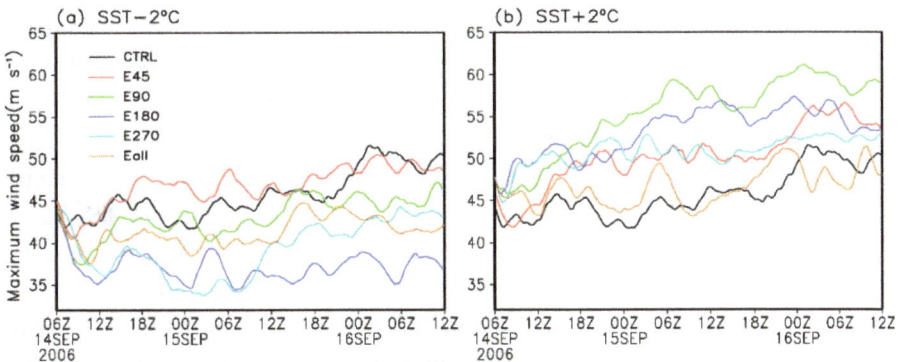

Fig. 9.2. Temporal evolutions of MSLP in the sensitivity experiments.

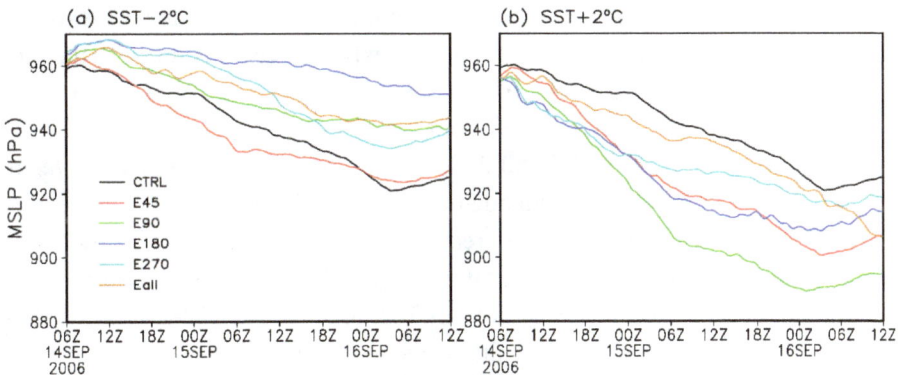

Fig. 9.3. Temporal evolutions of maximum wind speed at 10 m in the sensitivity experiments.

Comparing E45-2 with CTR, it can be seen that, the contribution of the SST in the eye region to the storm intensity is relatively small. This is consistent with results of XW10 when the surface enthalpy flux is changed in the eye region in one of their experiments. In E45+2, however, the increase of SST and thus the surface enthalpy flux in the eye can reduce RMW (Table 9.1). Similar results are found in Table 9.1 of XW10. The region with a radius of 45 km contains not only the eye, but also part of the eyewall in E45+2, suggesting that the increase of the surface enthalpy flux inside the eyewall contributes to the significant increase of TC intensity and the additional reduction of RMW compared with that of CTR (Figs. 9.2b, 9.3b and Table 9.1). Therefore, there may exist a positive feedback between the increase of the surface enthalpy flux inside the eyewall and the reduction of RMW in E45+2. The smaller increase of SST in E45+1, however, cannot trigger this positive feedback mechanism, and thus results in a negligible change of RMW and intensification of TC.

As shown in Figs. 9.2 and 9.3, E180-2 and E90+2 are the experiments with the most notable decrease and increase of TC intensity, respectively. Compared with the CTR, the storm minimum MSLP is increased by 30.1 hPa in E180-2 and decreased by 31.3 hPa in E90+2, and the maximum wind speed is reduced by 7.7 $m \ s^{-1}$ for a 2°C decrease of inner SST in E180-2 and increased by 8.5 $m \ s^{-1}$ for a 2°C increase in E90+2 (Table 9.1). Based on above results, we define an effective radius (ER) of about 1.5–2.0 times the RMW (i.e., not the RMW in CTR, but the RMW in each experiment), within which the SST is effective for the maintenance and intensification of the TC. SST within the ER is defined as "inner SST" in this study while outside the ER it is named "outer SST". The size of the ER is basically consistent with that suggested by XW10 and Sun *et al.* [2013a], but different from that of Miyamoto and Takemi [2010], who indicated that the effective radius of enthalpy flux is about 7-8 times the RMW.

The experiments with SST being changed farther outside of the eyewall, e.g., E270 and Eall, can be viewed as the outer SST experiments in contrast to the inner SST experiments (e.g., E90+2 and E180-2). In other words, E270 and Eall are the sensitivity experiments with outer SST being changed if we consider the inner SST experiments (e.g., E90+2 and E180-2) as references. As mentioned above, the outer SST is defined as the SST outside the ER. It can be seen that the decrease of SST in E270-2 does not cause such a great magnitude decrease as in E180-2 (Figs. 9.2 and 9.3). Similarly, the increase of SST in E270+2 doesn't cause the same magnitude increase of TC intensity as in E90+2. Results of E270-2 versus E180-2 and E270+2 versus E90+2 suggest that a decrease of outer SST suppresses the weakening of the simulated storm, while an increase of outer SST suppresses the intensification of the simulated storm. The same conclusion can be obtained by comparing Eall-2 with E180-2 and Eall+2 with E90+2. Above results are largely attributed to the fact that the decrease of outer SST weakens the activity of outer spiral rainbands while the increase of outer SST reinforces it. As suggested by Wang [2009] and Hill and Lackmann [2009], the TC intensity will reduce as convection in the outer spiral rainbands intensifies and the opposite is true when the convection weakens there. Thereby, the outer SST is responsible for the less change in TC intensity in the entire-SST experiments (e.g., Eall-2 and Eall+2) than that in the inner SST experiments (e.g., E180-2 and E90+2), namely, TC intensity is relatively insensitive to the entire SST change.

The TC intensity change is not only related to the extent and magnitude of SST changes but also associated with the mean SST itself. It is found that 30°C may be a critical temperature for the development of a TC, since the TC experiences rapid intensification when the SST is between 27°C and 30°C but slows down as the SST is above 30°C [Chan *et al.*, 2001]. SST near the TC in CTR is basically between 29°C and 32°C. Comparing Eall+2 with the CTR, the TC intensity is not increased notably (except for the last 6 hours in Eall+2 in Fig. 9.2(b)), following the increase of the entire SST. The outer SST is related to the mean SST and may be responsible for the insensitivity of TC intensity to the entire SST increase under warm SST conditions, for the difference of TC intensity between Eall-2 and E180-2 is much smaller than that between Eall+2 and E90+2 (Figs. 9.2 and 9.3). It is concluded that the outer SST contributes little to storm intensification when the mean SST is relatively cold (see the difference between Eall-2 and E180-2 in Figs. 9.2 and 9.3), whereas it plays a critical role in reducing TC intensity when the mean SST is larger than 30°C (see the difference between Eall+2 and E90+2 in Figs. 9.2 and 9.3).

The above results imply that the TC may not become much stronger as the entire SST increases since the opposite effect of SST inside and outside the ER cancels each other, resulting in little changes in TC intensity. Whether TCs will become stronger as the entire SST increases under the global warming condition in future, however, remains a controversial issue and requires further investigation.

9.3.2 *Storm inner-core size*

To provide a detailed description of the eyewall radius, the evolution of the overall inner-core size of the simulated storms is defined as the azimuthal mean RMW at 10 m (Fig. 9.4). Since the eyewall of the simulated storm is asymmetric from time to time [Wang, 2007], we only use the azimuthal mean as a proxy of the location of the overall eyewall. As is shown, the SST changes over different radial extents play radically different roles in determining the storm inner-core size.

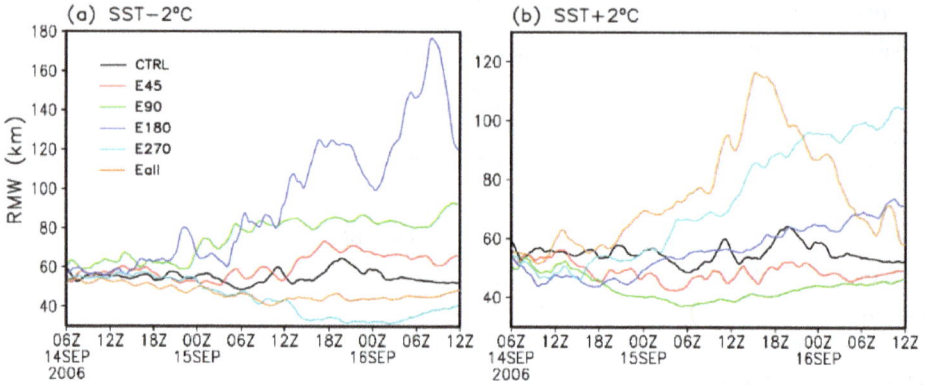

Fig. 9.4. Temporal evolutions of RMW at 10 m in the sensitivity experiments.

Due to the decrease of SST in the eye region, the simulated RMW in E45 is notably increased as compared to that in the CTR. The opposite is true as SST increases in the eye region. However, the decrease of inner SST within the ER (such as in E180-2) is responsible for the greatest increase in the inner-core size of TC after 12-h model spin-up time. Similarly, the increase of inner SST within the ER (such as in E90+2) reduces the TC inner-core size. This is somewhat different from the idealized model results of XW10, which ignored the impact of the surface enthalpy flux near the eyewall on the TC inner-core size. On the contrary, it is found that the decrease of outer SST is responsible for the sharp decrease of the storm inner-core size while the increase of outer SST results in a sharp increase of it. This is clearly reflected in the difference between results of Eall-2 and E180-2 and that between results of Eall+2 and E90+2 (Fig. 9.4). The strong impact of outer SST on TC inner-core size indicates that TC may become much larger as the entire SST increases.

Interestingly, there is a notable negative correlation between TC intensity and its inner-core size. Table 9.1 shows the differences between the sensitivity experiments and CTR. These differences demonstrate various impacts of SST change over different radial extents on TC intensity and its inner-core size. Based on the statistical analysis of observations of many storms, Stern and Nolan [2009] suggested that the size of the RMW is directly proportional to the outward slope of the RMW with height, but there is no distinct relationship between the size of the RMW and TC intensity. However, they also emphasized that for a given simulation, the RMW is approximately inversely proportional to TC intensity. Their results are consistent with the findings of Emanuel [1986], who suggested that for the simulation of a single storm, the storm intensity appears to be closely related to slope of the RMW and thus its size, although the relationship is nonlinear. Results in our study reveal a strong relationship between the RMW and TC intensity, despite the fact that the underlying SST is varied. Details will be further discussed in section 4.2. According to this negative correlation, the dramatic decrease of RMW in Eall+2 may be responsible for the increased intensification of TC after 0300 UTC 16 September 2006. It is noteworthy that the storm intensity in Eall+2 is still much weaker than that in E90+2 even after its increased intensification.

9.4 Possible reasons for the change of TC intensity

9.4.1 *Impact of SST through surface enthalpy flux*

9.4.1.1 *Area-integrated surface enthalpy flux*

Previous studies suggested that a TC intensifies and maintains its strength against surface frictional dissipation by extracting energy from the underlying oceans. Thus, energy exchange at the air-sea interface is the key to the intensity change of a TC [Malkus and Riehl, 1960; Black and Holland, 1995]. Fig. 9.5 shows the temporal evolution of area-integrated kinetic energy at 10 m and the surface enthalpy flux within a 200-km radius in some representative experiments. Owing to the effect of warm SST and other favorable conditions, the area-integrated surface enthalpy flux and kinetic energy increase from 0600 UTC 14 September to 0300 UTC 16 September. Note that a significant reduction in the surface enthalpy flux and kinetic energy occurs at about 0300 UTC 16 September 2006 in these sensitivity experiments except for Eall-2, when the simulated TC passed over a cold pool near Miyako Island during that period.

In the experiments, though the TC intensity in E90+2 is much stronger than that in Eall+2 (Fig. 9.2), the integrated surface enthalpy flux and kinetic energy in E90+2 are notably less than those in Eall+2 (Fig. 9.5). In contrast, the TC intensity in E180-2 is much weaker than that in Eall-2 (Fig. 9.3), but the integrated surface enthalpy flux and kinetic energy in E180-2 are just slightly higher than those in Eall-2 (Fig. 9.5). Apparently differences in the integrated surface enthalpy flux might explain the difference in storm intensity between sensitivity experiments and CTR, but cannot explain the difference between the inner SST experiments (e.g., E180-2 and E90+2) and the entire SST experiments (e.g., Eall-2 and Eall+2). There must have other factors that promote or suppress storm intensification in the entire SST experiments. We will discuss this issue later.

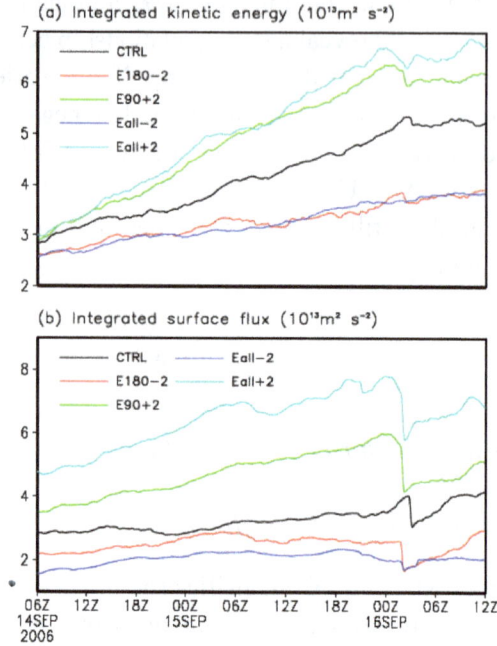

Fig. 9.5. (a) Temporal evolution of area-integrated kinetic energy at 10 m; (b) area-integrated the surface enthalpy flux (calculated within a radius of 200 km).

9.4.1.2 *Different effects of the surface enthalpy flux inside and outside the ER*

The wind-induced surface heat exchange, which describes a positive feedback between the increase in the surface enthalpy flux and the surface wind speed in the near-core region of a TC, is regarded as the dominant process that controls the rapid intensification of a TC [Emanuel, 1986; Rotunno and Emanuel, 1987]. Hence the surface enthalpy flux may play a more important role in determining TC intensity in the near-core region than in other regions. As mentioned above, the SST and thus the surface enthalpy flux within the ER contributes greatly to the storm intensification, while the surface enthalpy flux outside the ER actually reduces the storm intensity, suggesting that the effects of SST anomalies over different radial extents on TC intensification are substantially different.

Fig. 9.6 illustrates the distributions of the surface enthalpy flux difference and reflectivity difference between two specific experiments (i.e., E180-2 and E90+2) and CTR, averaged over the mature stage (from 1800 UTC 15 September to 0600 UTC 16 September). It clearly shows that the surface enthalpy flux near the eyewall reduces considerably with the decrease of inner SST, leading to weaker convection near the eyewall in terms of simulated reflectivity and thus weaker TC intensity. Meanwhile, the decrease of inner SST inside the ER also induces an increase in convective activity of outer spiral rainbands. The opposite happen when the inner SST is increased. As proposed by Houze *et al.* [2006] and Houze *et al.* [2007], there is a competition between effects of convection in outer spiral bands and that near the eyewall. This competition plays a vital role in determining the change of TC intensity. Wang [2009] has suggested that active convection in outer spiral rainbands can reduce the storm intensity. Thus, the increase in inner SST contributes greatly to TC intensification by enhancing convection near the eyewall and weakening the convection in outer spiral rainbands. The decrease in inner SST has the opposite effects.

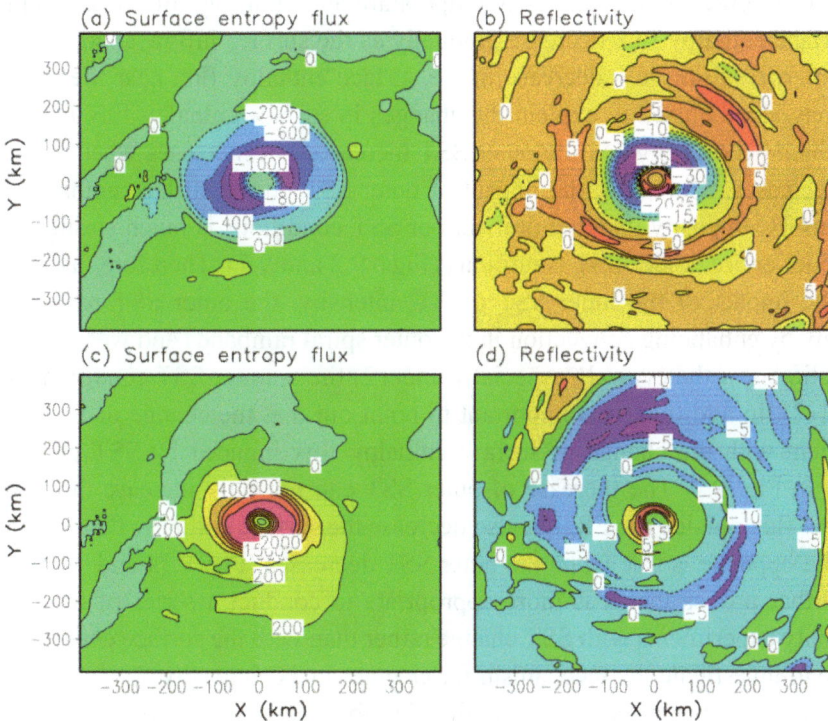

Fig. 9.6. Distributions of the surface enthalpy flux difference (W m^{-2}) and max-reflectivity difference (dBZ) between results of sensitive experiments and CTR. (a) Surface enthalpy flux difference between inner SST experiments E180-2 and CTR; (b) Max-reflectivity difference between inner SST experiments E180-2 and CTR; (c) Surface enthalpy flux difference between E90+2 and CTR; (d) Max-reflectivity difference between E90+2 and CTR. (All calculation is done over the mature stage from 1800UTC on 15 September to 0600UTC on 16 September).

To investigate the impact of the outer SST on TC intensity, we compare the distributions of the surface enthalpy flux and reflectivity between the inner SST experiments (e.g., E180-2 and E90+2) and the entire SST experiments (e.g., Eall-2 and Eall+2, Fig. 9.7). The difference between E180-2 and Eall-2 can be regarded as the result of the increased outer SST, while the difference between E90+2 and Eall+2 reflects the impact of the decreased outer SST. It is found that the increase in outer SST not only induces more surface enthalpy flux outside the ER and thus stronger outer spiral rainbands, but also reduces the surface enthalpy flux inside the ER and thus the activity of convection near the eyewall (Fig. 9.7). Above results are completely reversed when the outer SST decreases. Impacts of the increase (decrease) in outer SST are similar to that of the decrease (increase) in inner SST (Figs. 9.6 and 9.7). As suggested by Sun *et al.* [2013a], due to the impact of the warm outer SST, the air-sea temperature and humidity difference and thus the surface enthalpy flux decrease gradually as the surface inflow approaches the TC center, resulting in a decrease in the surface enthalpy flux near the eyewall. This may also be used to explain the changes in surface enthalpy flux inside the ER, which decreases when the outer SST increases but increases when the outer SST decreases. Moreover, the impact of outer SST on the surface enthalpy flux and thus TC intensity under entire warmer SST conditions is much stronger than that under entire colder SST conditions (Figs. 9.2 and 9.7). Therefore, in contrast with the impact of the inner SST on TC intensity, the outer SST reduces TC intensity by enhancing convection in the outer spiral rainbands and weakening the convection near the eyewall, especially under entire warmer SST conditions (e.g., in Eall+2). In addition, it is important to point out that the change of SST is not consistent with the change of surface enthalpy flux induced by SST change as shown in Fig. 9.7. The increase of outer SST can not only increase the surface enthalpy flux outside the ER, but also decrease the surface enthalpy flux inside the ER. The opposite is true when the outer SST decreases (Fig. 9.7a and 9.7c). This fact further proves that it is more appropriate to conduct a series of reasonable sensitivity experiments with SST change rather than with the surface enthalpy flux change to investigate the mechanism for changes in TC intensity.

Changes in the surface enthalpy flux distribution have a significant impact on the storm inner-core size by influencing the activity of outer spiral rainbands. Comparing E180-2 with Eall-2, the warmer outer SST in E180-2 results in larger amount of simulated surface enthalpy flux released outside the ER, which is favorable for the development of outer spiral rainbands (Figs. 9.7a and 9.7b). The pressure outside the RMW and hence the pressure gradient across the RMW decreases subsequently. These changes help weaken the storm but increase the storm inner-core size. All above changes and their impact on the storm intensity and the inner-core size are reversed when the outer SST decreases, which shows the impact of colder outer SST in E90+2 compared to that in Eall+2 (Figs. 9.7c and 9.7d). Note that the effect of the inner SST decrease (increase) is similar to that of outer SST increase (decrease) outside the ER. Above results clearly indicate that the inner and outer SST changes can either increase or decrease the activity of outer spiral bands, leading to changes in both the storm intensity and the storm inner-core size (Table 9.1).

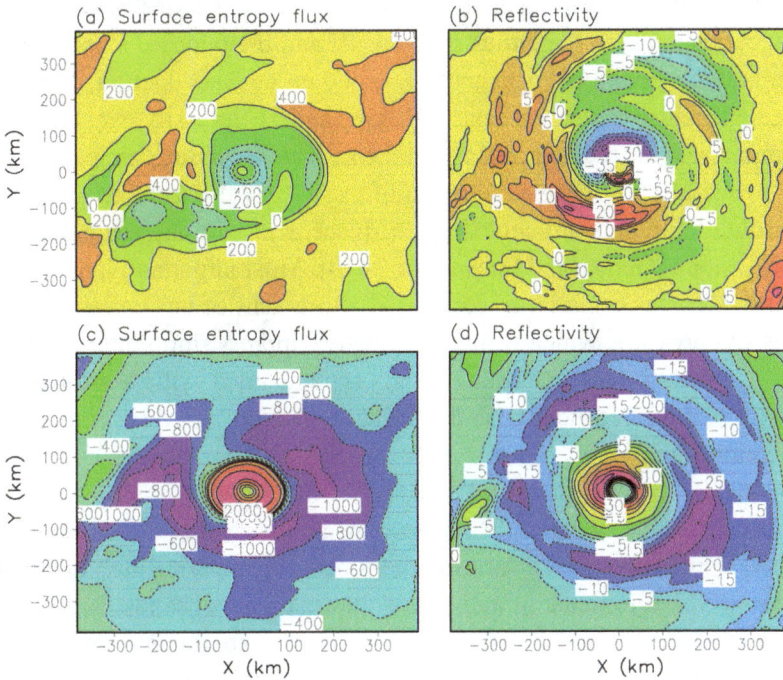

Fig. 9.7. As in Fig. 9.6, but for the difference between E180-2 and Eall-2 (a, b), and the difference between E90+2 and Eall+2 (c, d).

The effect of the entire SST change on TC intensity and inner-core size can be regarded as the sum of the effects of inner and outer SST. As the entire SST increases, the effects of the inner SST that help intensify the TC are stronger than the effects of the outer SST that suppress the TC intensification. Meanwhile, the effects of the inner SST that reduce the inner-core size are weaker than the effects of the outer SST that increase the inner-core size. As a result, the TC becomes stronger and larger as the entire SST increases and the opposite effects happen when the entire SST decreases. However, the positive effect of the entire SST on TC intensification is small under warm area-averaged SST condition. This is consistent with the findings of Sun *et al.* [2013a].

9.4.2 *Impact of SST in view of the thermodynamic structure*

9.4.2.1 *Latent heating*

Many recent studies suggested that energy exchange between the ocean and atmosphere affects TC structure [e.g., Chen *et al.*, 2010; Lee and Chen, 2012]. In the previous section, we have shown a negative correlation between TC intensity and inner-core size when imposing SST changes at various radial extents. Stern and Nolan [2009] found a linear relationship between the size of RMW (which basically corresponds to the radius of eyewall) and the outward slope of eyewall based on both observational data analysis and theoretical deduction. They argued that the relationship between the eyewall slope and TC intensity may be attributed to the impact of the radial pressure gradient. The ascending air parcels in the eyewall normally experience a reduction of the inward-directed pressure gradient force, leading to an outward centrifugal displacement with increasing height, i.e. the eyewall slope [Gentry and Lackmann, 2010]. Most importantly, a less upright eyewall often comes out with a weaker radial gradient of tangential and vertical velocities and pressure, which would eventually suppress storm intensification. In contrast, a more upright eyewall is often accompanied with a stronger radial gradient of vertical velocity and pressure, which would promote storm intensification [Fierro *et al.*, 2009; Sun *et al.*, 2013b]. Thereby, the TC intensity is related to the TC eyewall slope and thus the TC inner-core size.

In this study, the change of RMW may be an important index to estimate the change of TC intensity as SST varies in the sensitivity experiments. As mentioned above, due to the difference in activity of outer spiral rainbands caused by SST difference, the simulated storms in the sensitivity experiments exhibit a large difference in inner-core size. The storm inner-core size determines the radius at which strong latent heating occurs, and thus contributes to storm intensification.

Latent heating is the direct forcing that drives the eyewall convection and the secondary circulation of TC, which makes the storm to maintain its strength and develop further. In this study, the latent heating is found to be sensitive to the change of SST and its radial extent. This is mainly because the SST-related surface enthalpy flux is favorable for active convection in the storm, which will increase the latent heat release [Wang, 2009]. As suggested by Eliassen [1951] and Holland and Merrill [1984], one of the fundamental conditions for TC development is that heating occurs in the region of high inertial stability. The heating in the high-inertial-stability region inside the RMW is efficient at generating a localized temperature tendency, and this efficiency increases dramatically with storm intensity; whereas the heating in the low-inertial-stability region outside the RMW is inefficient at generating a warm core. In other words, the vortex intensification rate depends on how much of the heating is occurring inside the RMW. Moreover, Vigh and Schubert [2009] proposed that heating in the outer region of about 2-3 time (80 km: 30 km) the RMW from TC center could still spin up the local tangential wind and radially constrained the circulation response, though the heating is occurring outside of the RMW.

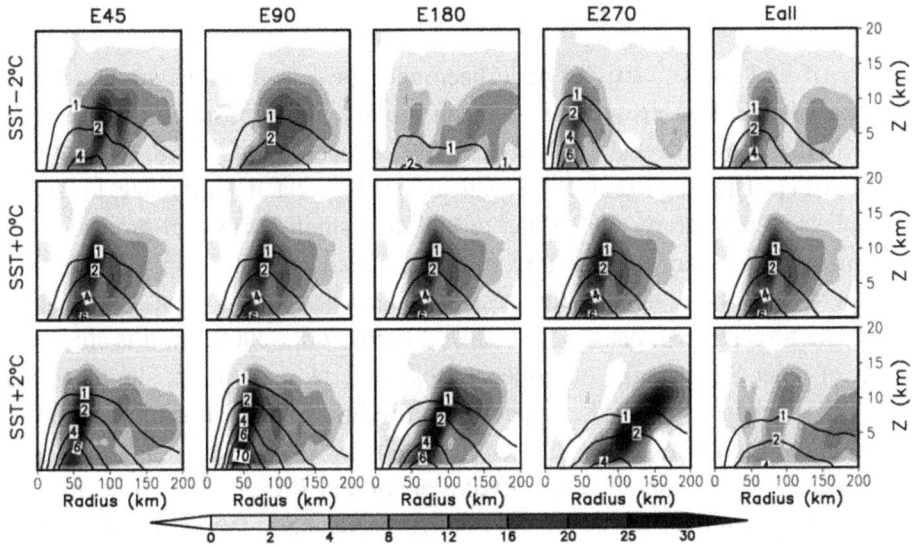

Fig. 9.8. Azimuthal- and time-averaged cross sections of the model-simulated latent heat rate (°C h⁻¹; shaded) and radial pressure gradient (10^{-2} Pa m⁻¹; contoured) in the TC mature stage.

Fig. 9.8 shows the azimuthal- and time-averaged cross sections of the model-simulated latent heating at the TC mature stage. As expected, due to more energy exchange at the air-sea interface (i.e., the surface enthalpy flux), all the warmer-SST experiments except for Eall+2 produce larger magnitudes of latent heating near the eyewall. The larger latent heating results in a larger central pressure fall and thus larger radial pressure gradient. However, compared with results in the entire warmer-SST experiment (i.e., Eall+2), the simulated latent heating in the inner warmer-SST experiments are concentrated in much smaller areas close to the eye due to the smaller RMW and eyewall slope. Similar results are found when we compare simulations of the entire colder-SST (i.e., Eall-2) experiment with that of the inner colder-SST experiments. Hack and Schubert [1986] have shown that the smaller the radius at which latent heating occurs, the more significant contribution it brings to the central pressure fall and radial pressure gradient near the eyewall. Hence the simulated latent heating at the smaller radius in the inner warmer-SST experiments and the entire colder-SST experiment contributes to a deeper central pressure fall and larger pressure gradient, and eventually leads to either a larger increase or a smaller decrease in TC intensity (Figs. 9.2 and 9.3). Conversely, the larger radius in the inner colder-SST experiments and entire warmer-SST experiment is responsible for the larger decrease or smaller increase in TC intensity (Figs. 9.2 and 9.3).

9.4.2.2 *Upper-tropospheric temperature*

Impacts of SST on TC activity have been addressed in previous studies with a focus on whether the changes of SST can modify the tropospheric temperature profile in the tropics [Vecchi and Soden, 2007; Vecchi *et al.*, 2008; Ramsay and Sobel, 2011]. The large-scale SST can affect TCs via its influence in upper-tropospheric temperature, which has to maintain small horizontal gradients and thus must adjust to stay consistent with the SST changes in the tropics. Moreover, the upper-tropospheric temperature responds much less to the local SST than the large-scale SST [Vecchi and Soden, 2007; Vecchi *et al.*, 2008]. In this study, the impact of inner SST change on upper-tropospheric temperature is similar to that of local SST change, while the impact of entire SST change is similar to that of large-scale SST change.

Fig. 9.9 shows the temporal evolution of the difference in the domain-averaged (average over the entire D3) upper-tropospheric temperature at 15-km height (approximately the height of outflow layer) between the sensitivity experiments and CTR. Comparing with that of CTR, as the extent of SST anomalies increases in the sensitivity experiments, the change of upper-tropospheric temperature will increase notably. Apparently the change of upper-tropospheric temperature is sensitive to the extent of SST change. Compared with the experiments with the change of entire SST, the defined inner SST change contributes less to the change in upper-tropospheric temperature. Thereby, the increase of inner SST causes a much less increase of upper-tropospheric temperature than that induced by the increase of the entire SST, which results in a larger difference between sea surface air temperature and upper-tropospheric temperature and thus more unstable atmospheric condition. This unstable condition favors the intensification of the TC. The opposite happen as the inner SST decreases. In contrast, analysis of the difference in upper-tropospheric temperature between the inner-SST experiments (e.g., E90+2 or E180-2) and entire-SST experiments (e.g., Eall+2 or Eall-2) shows that the outer SST plays a significant role in determining upper-tropospheric temperature (Fig. 9.9). Increase in outer SST makes the local stratification more stable and thus suppresses the intensification of TC. Note that, different from the concerns in the previous studies [e.g., Vecchi and Soden, 2007; Vecchi *et al.*, 2008; Ramsay and Sobel, 2011], our concern is on "transient" response of upper-tropospheric temperature to the SST on storm scale, which depends on the strength of convective activity. Therefore, convective activity could affect upper-tropospheric temperature by influencing not only vertical flux of temperature but also condensation in the anvil clouds [Sun *et al.*, 2014a].

This study reveals the different impacts of inner and outer SST changes on upper-tropospheric temperature and explores the physical mechanisms of TC intensification. Due to the strong convection in TC region, however, the involved mechanisms discussed here may not be applicable to other tropical regions.

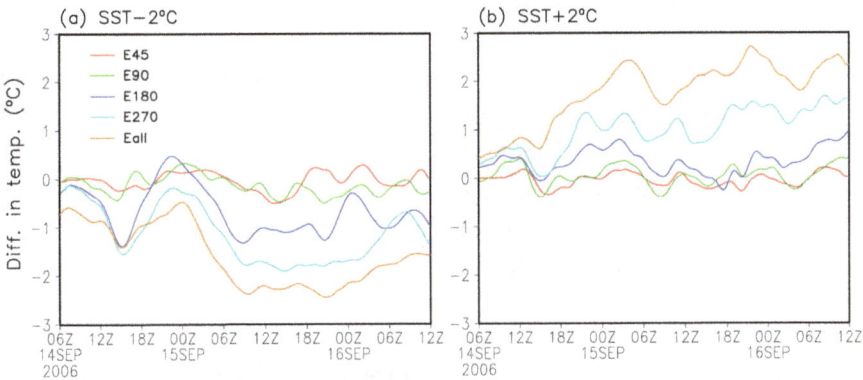

Fig. 9.9. Temporal evolutions of the difference in the averaged upper-tropospheric temperature (°C) at 15 km in the entire moving domain (D3) between sensitivity experiments and CTR.

9.4.3 *Validation on the opposite effects of inner and outer SST*

To verify and validate the effects of inner and outer SST on TC intensity, a suite of semi-idealized sensitivity experiments are conducted. While the imposed SST anomalies are unrealistic, results of these experiments will help us understand the physical and dynamic mechanisms of the SST impact on TC intensity, which is important for successful forecast of TC intensity and inner-core size especially when a TC comes across an ocean cold or warm pool. Previous studies suggested that TC intensity undergoes a significant change as the TC passes over an ocean warm or cold pool [e.g., Shay *et al.*, 2000; Lin *et al.*, 2005; Wu *et al.*, 2007]. Results of the present study indicate that TC intensity may start to change even before the TC passes over the ocean warm or cold pool.

To validate our conclusions on the opposite effects of inner and outer SST, we have conducted two additional sensitivity experiments to investigate the response of TC intensity to a fixed change of SST. The fixed SST anomalies is either reduced or increased by 2°C within a radius of 180 km in the sensitivity experiments (EF180). The center of the fixed anomalies is consistent with the TC center at 0000 UTC 16 September 2006 when the TC is approximately steady. Fig. 9.10 shows the temporal evolution of MSLP and RMW for the fixed SST anomalies experiments (i.e., EF180-2 and EF180+2) and CTR, and the differences between the fixed SST anomalies experiments and CTR. As a TC approaches a cold or warm pool, the impact of the pool on TC intensity is related to the distance from the TC center to the pool. Before 1800 UTC 14 September, the TC center is about 350 km away from the fixed ocean cold pool in EF180-2. Comparing EF180-2 and CTR, due to the long distance between the cold pool and the TC center, the impact of the SST anomalies on TC intensity is very slight and can be neglected. As the TC approaches the fixed cold pool but the cold pool is still outside the ER of the TC (approximately from 1800 UTC 14 to 1800 UTC 15 September), the impact of the cold pool on TC intensity becomes much stronger. Consistent with the aforementioned impact of the outer SST, the fixed cold pool outside the ER increases the TC intensity, but reduces the TC inner-core size (Figs. 9.10a and 9.10c). However, as the TC further approaches the fixed cold pool and the fixed cold pool is within the ER of the TC (after 1800 UTC 15 September), the cold pool reduces the TC intensity (Fig. 9.10a), but increases the TC inner-core size (Fig. 9.10c). Apparently, the impact of the cold pool within the ER of the TC is consistent with the aforementioned impact of the inner SST.

On the other hand, the impact of the fixed ocean warm pool on the TC intensity is notably weaker than that of the fixed ocean cold pool (Fig. 9.10a and 9.10b). This is consistent with the results of Chan *et al.* [2001], which suggested that the sensitivity of TC intensity to SST may decrease under higher SST condition (e.g., the SST is above 30°C). Despite its relatively small impact on the TC intensity, the fixed ocean warm pool contributes greatly to the increase of the TC inner-core size. As the TC moves closer to the fixed ocean warm pool but the warm pool is still outside the ER of the TC (from 0000 UTC 15 to 1800 UTC 15 September), the TC inner-core size increases notably due to the impact of the fixed ocean warm pool, which is consistent with the aforementioned the impact of outer SST. However, as the TC gets closer to the fixed ocean warm pool and the fixed warm pool is within the ER of the TC (after 1800 UTC 15 September), the TC inner-core size does not decrease notably as expected. This may be attributed to the large radius of the ocean warm pool (i.e., 180 km), which covers not only the region inside the ER but also the region outside the ER. Thus, the impact of the increase of inner SST on the TC inner-core size is offset by that of the increase of outer SST, resulting in a steady TC inner-core size after 0000 UTC 16 September.

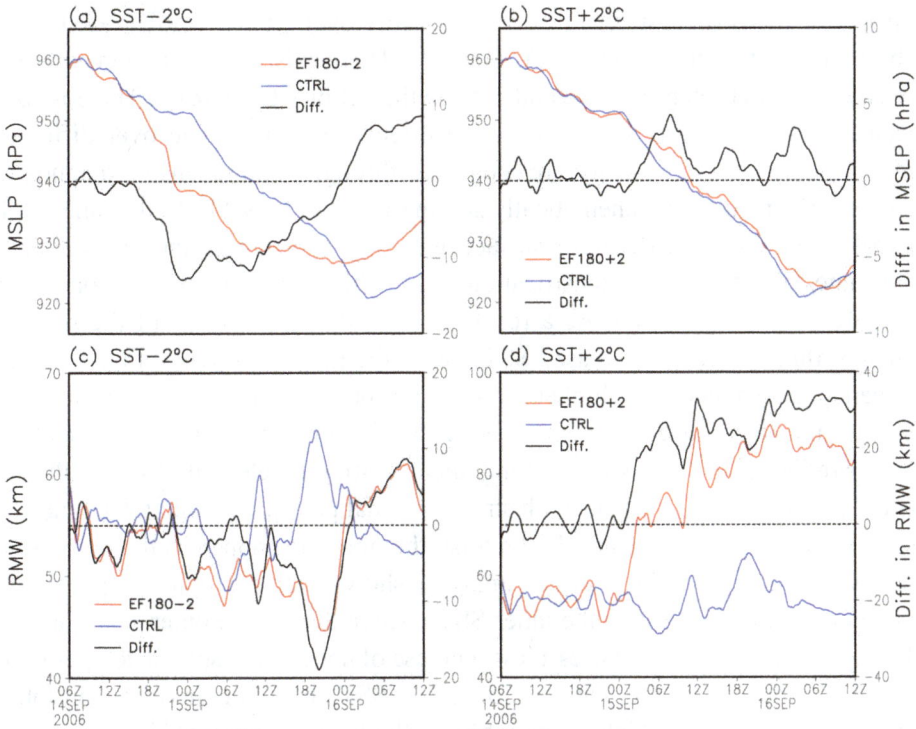

Fig. 9.10. Temporal evolutions of the MSLP and RMW for the fixed SST anomalies experiments (i.e., EF180-2 and EF180+2) and CTR, and the differences between the fixed SST anomalies experiments and CTR.

9.5 Discussions

The results in this study indicate that the SST in the TC eye region contributes little to the storm intensity. Instead, the inner SST within the range 1.5-2.0 times the RMW (defined as ER) plays a significant role in TC intensification and can reduce the inner-core size of the TC. In contrast, the outer SST contributes greatly to the growth of TC inner-core size, but reduces the TC intensity. Changes in the inner SST have stronger impact on TC intensification than changes in the entire SST, whose impact is weakened by the negative effect of outer SST on TC intensification.

The mechanisms for the different effects of SST over different radial extents on TC intensification are discussed from perspectives of the surface enthalpy flux and TC structures. Different from many previous studies that focus on the impact of the surface enthalpy flux, more attention was paid to the impact of SST on the storm structure. Our present study based on both the surface enthalpy flux and TC structure analysis suggests that the change of SST over different radial extents has substantially different impacts on TC intensity. There exists a competition between the effects of SST change inside and outside the ER on TC intensity. The possible mechanisms responsible for the different effects of SST change over different radial extents on TC intensity are identified through comprehensive diagnostic analysis. The results are schematically summarized in Fig. 9.11. As the inner SST increases within the ER, more surface enthalpy flux enters the eyewall and contributes directly to TC intensification. Meanwhile, the impact of SST on storm structure is also significant. As a result of the radial distribution of the surface enthalpy flux caused by increased SST within the ER, the activity of outer spiral rainbands is weakened. Such changes not only promote storm intensity but also decrease the storm inner-core size. The reduced storm inner-core size leads to a smaller radius of eyewall where strong latent heating is released. As a result, the central pressure of TC deepens with stronger radial pressure gradient, leading to a larger increase in TC intensity. In contrast, the opposite happen as the outer SST increases outside the ER. Further analysis shows that the upper-tropospheric temperature responds less to the inner SST change than to the outer SST change. The increase of inner SST causes a less increase of upper-tropospheric temperature than that induced by the increase of outer SST, resulting in more unstable atmospheric condition which is favorable for the intensification of TC.

In this study, the positive effects of the inner SST increase within the ER are strong enough to overcome the negative effects of the outer SST increase and result in a stronger TC. Due to the negative effects of the outer SST, however, the storm intensity simulated by the experiments with SST increase over the entire domain is notably weaker than that in the warmer inner SST experiments. Similarly, the opposite effects of inner SST and outer SST increase can also be used to explain the response of TC intensity to the decrease of the SST inside and outside the ER. How TC intensity responds to the change of entire SST depends on the competitive effects of SST changes inside and outside the ER.

In addition, the competition between the effects of SST changes inside and outside the ER and thus the change of TC intensity in the entire SST experiments is related to the area-averaged SST. The TC intensification caused by the entire SST increase under warmer SST conditions is notably weaker than that under colder SST conditions. While several previous studies have suggested that TCs will become much stronger when the entire SST increases, results of our study indicate that this is not always the case. One possible reason for the discrepancy is because the negative effect of outer SST on TC intensity is largely underestimated in previous studies. While response of TC intensity to the increase of entire SST in the context of global warming still remains controversial, results of this study will help shed light on this issue and reduce the uncertainties in prediction of future change scenario.

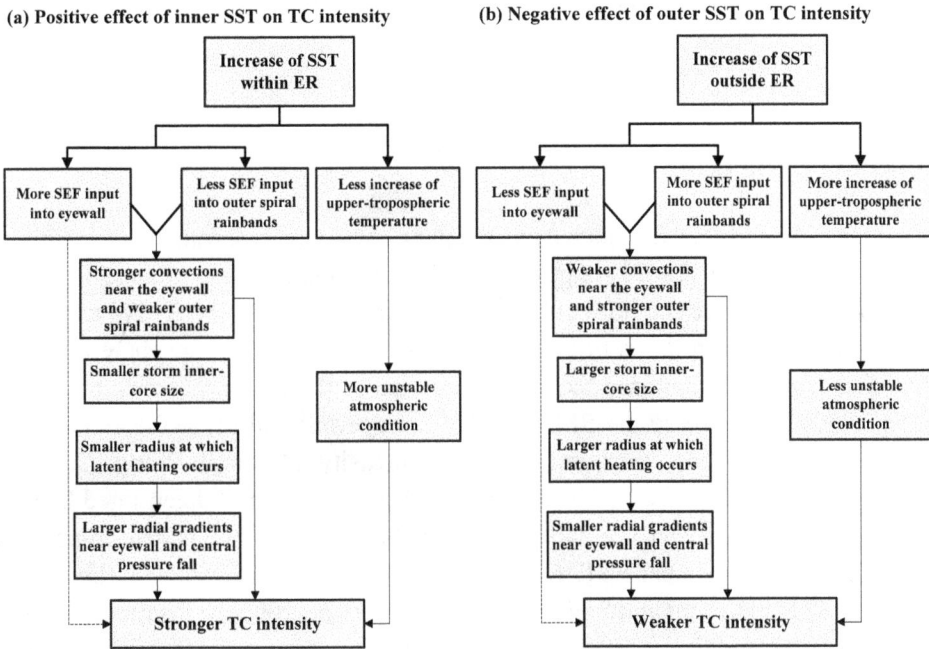

Fig. 9.11. Schematic diagram summarizing the possible mechanisms responsible for different effects of SST change over different radial extent on TC intensity. The thin dashed (thin solid) line indicates the impact of SST on TC intensity in view of the surface enthalpy flux (storm structure).

Furthermore, understanding the mechanisms for the opposite effects of inner and outer SST is also vital to the forecast of variations in TC intensity and inner-core size when a TC comes across an ocean cold or warm pool. As the TC approaches a cold or warm pool, the impact of the pool on TC intensity is related to the distance from the TC center to the pool. As the ocean cold pool is outside the ER from TC center, similar to the impact of inner SST anomaly, the ocean cold pool increases the TC intensity but reduces the TC inner-core size; while as the cold pool is within the ER from the TC center, similar to the impact of outer SST anomaly, it reduces the TC intensity but increases the TC inner-core size. The opposite phenomena occur when the TC comes across an ocean warm pool.

Chapter 10

Effects of Relative and Absolute SST on Tropical Cyclone Intensity

10.1 Introduction

It has been found that the remote energy input plays a negative role in determining TC intensity. Wang [2009] suggested that heating due to phase change in outer rainbands will reduce the horizontal pressure gradient across the RMW and thus the storm intensity in terms of maximum wind in the lower troposphere and, on the other hand, will increase the inner-core size of the storm. Moreover, strong outer winds can increase the SEFs outside the eyewall and favor the activity of outer spiral rainbands, thus reducing TC intensity [Xu and Wang, 2010b]. Furthermore, many studies have shown that TC intensity is not determined by the absolute value of the SST (hereafter "absolute SST"), instead, it is determined by the relative value of local SST, relative to the spatially-averaged SST (hereafter "relative SST") [Vecchi and Soden, 2007; Vecchi et al., 2008; Ramsay and Sobel, 2011]. Ramsay and Sobel [2011] suggested that, the computed potential intensity (PI) is more sensitive to relative SST than to absolute SST, with slopes of between about 7 and 8 m s^{-1} °C^{-1} in their relative SST experiment and about 1 m s^{-1} °C^{-1} in their absolute SST experiment. Furthermore, previous studies focused on the effect of the radial extent within which the underlying SST contributes greatly to TC intensity by supplying SEF into the TC, though this is still a point of scientific debate. Miyamoto and Takemi [2010] proposed that the important radial extent of a TC is about 7–8 times the RMW, while the radial extent suggested by recent studies [Xu and Wang, 2010a; Wang and Xu, 2010] is only about 2–2.5 times RMW.

About the importance of relative SST on TC intensity, it has been pointed out that the PI of a TC is sensitive to both the local SST and the local tropospheric temperature profile, and the local tropical upper-tropospheric temperature is controlled by the tropical mean SST [Emanuel, 1986; Holland, 1997; Bister and Emanuel, 1998]. A warming (cooling) relative SST acts to warm and moisten (cool and dry) the atmospheric boundary layer locally, but does not cause changes of similar magnitude in the free-troposphere temperature since the latter must remain approximately uniformly horizontal. Such a warming (cooling) thus destabilizes (stabilizes) the overlying atmosphere, altering the PI [Vecchi and Soden, 2007]. Using a single-column model, Ramsay and Sobel [2011] suggested that the greater sensitivity of PI to relative SST can be attributed to a greater rate of increase in the air-sea thermodynamic disequilibrium with increasing SST, namely, the change in air-sea disequilibrium for a given SST change is smaller in absolute SST experiment than in relative SST experiment, reducing the slope of PI. However, due to the limitation of the single-column model in Ramsay and Sobel [2011], they could not provide the convincing evidence which supports the physical processes on the impact of relative SST on TC intensity.

To investigate how TC intensity responds to changes in relative SST and absolute SST and which physical processes are involved, we take the western Pacific TC Shanshan (2006) as an example and conduct two sets of experiments. The first set can be considered as a set of relative SST experiments which represent SST change in a limited radial extent while the rest of the SST remains unchanged, whereas the second set is a set of absolute SST experiments which changes the SST over the whole simulation domain uniformly. Although the design of relative and absolute SST experiments is basically similar to that in Ramsay and Sobel [2011], our particular focus is on the dynamic and thermodynamic processes associated with the response of SEF to given SST. Moreover, although in reality the SST distribution depends on the surface wind stress and the movement of ocean currents, understanding how a TC responds to its underlying SST is a critical step in improving our conceptual model of the coupled ocean-TC system.

10.2 Experimental design

The model used in this section is the WRF V3.3.1 [Skamarock *et al.*, 2008]. The model domain is triply-nested, with two-way interactive nesting and with the two inner meshes automatically moving to follow the model storm with the finest grid mesh at 5 km (see Fig. 9.1). The model integration started at 0000 UTC 14 September and ended at 1200 UTC 16 September, a total of 60-h integration before and after the recurvature of Shanshan (2006) (referred to as CTR). The initial and lateral boundary conditions are from the NCEP final analysis data (FNL). The model output is at 5-min intervals. In order to take the upwelling and feedback of the ocean on TC intensity into consideration [Chan *et al.*, 2001; Zhong and Zhang, 2006], the SST was from the TRMM Microwave Imager (TMI) level-1 standard product with 0.25°×0.25° resolution. More details on model configuration and simulation design are available in Sun *et al.* [2012], it was found that the track and intensity of Shanshan, as well as the structures of the eye, eyewall, spiral rainbands and other inner-core features can be reproduced reasonably well, and the time-averaged simulated track bias is about 55 km, the bias of MSLP is 4.7 hPa, and that of maximum wind speed is 4.4 m s^{-1} [Sun *et al.*, 2012]. The observed and simulated track of Shanshan at 6-h intervals is also shown in Fig. 9.1b.

Besides the CTR, 11 sensitivity experiments were conducted. In each sensitivity experiment, the underlying SST is either reduced or increased, within a specified radius from the TC center. Specifically, the change of SST is modified with a weighting function that is maximized in the TC center, changes linearly outwards to zero at a prescribed radius and remains zero further outwards, and thus the distribution of SST change is mainly consistent with that of an SST anomaly in the real situation.

Table 10.1 lists the sensitivity experiments and the maximum intensity and RMW of Shanshan in each experiment.

10.3 Effect of SST changes at different radial extents on TC intensity

It has been pointed out that the SEF in different regions plays a different role in determining TCs [Xu and Wang, 2010a; Wang and Xu, 2010]. However, SEF is determined not only by SST but also outer wind fields [Xu and Wang, 2010b], thus the role of SEF and SST on TC intensity may be different. Here, based on a series of sensitivity experiments, the different effects of SST changes at different radial extents on TC intensity are discussed.

Due to the difference of SST change at different radial extents, the intensity, as well as the inner-core size, defined by the RMW at 10 m above the sea surface, of TCs in different experiments are radically different. As shown in Table 10.1, the simulated storm intensity and the inner-core size in E45 slightly changed with SST in the eye region compared to that in CTR, which is also consistent with the SEF experiment by Xu and Wang [2010a].

In our experiments, the decrease (increase) of SST within a radius of about 180 km (90 km) is responsible for the most significant storm intensity decrease (increase), whereas 180 km (90 km) is about 2–3 times RMW in E180M (E90P) (Table 10.1). Since SST in this region for storm intensity is very important, it was called the critical region (CR). Note that the definition of the CR is similar to that suggested by Xu and Wang [2010a] and Wang and Xu [2010], but different from that of Miyamoto and Takemi [2010], which indicated that the SEF within the radius of about 7–8 times RMW plays a vital role in determining storm intensity.

It was found that the remote SST can also influence storm PI through its influence on upper atmospheric temperatures [Vecchi and Soden, 2007]. The remote SEF is crucial to the growth of storm inner-core size, but could reduce storm intensity [Xu and Wang, 2010a]. On the other hand, the remote SST can be taken as a diabatic heating to outer spiral rainbands, as Wang [2009] suggested, it may limit the TC intensity but increase the TC size. By comparing E270M (E270P) with E180M (E90P) in Table 10.1, one can see that the decrease (increase) of SST within a 270-km radius does not cause a great change in storm intensity as in E180M (E90P), but could decrease (increase) the inner-core size notably; and similar conclusions could be reached by comparing EallM (EallP) with E180M (E90P). This suggests that the decrease (increase) in remote SST outside CR is crucial to the reduction (growth) of the TC inner-core size, but could increase (reduce) TC intensity, which will lead to a decrease of sensitivity in the TC intensity to absolute SST change, especially under the warm SST condition such as SST $\geq 30°C$ in EallP. Next, we will provide an explanation for the impact of relative and absolute SST on TC intensity.

Table 10.1. Summary of the control experiment and sensitivity experiments, where r is the radial distance from the storm center, ΔSST is the modification of SST, $MSLP_{min}$ and V_{max} are the lifetime minimum MSLP and lifetime maximum wind speed at 10 m for each simulation, respectively. RMW_{ave} is the RMW at the 10 m averaged in the intensification stage (from 0000 UTC 15 September to 0600 UTC 16 September) when the difference between the sensitivity experiment and CTR is significant.

Exp.	Modification to the underlying SST	$MSLP_{min}$ (hPa)	V_{max} (m s⁻¹)	RMW (km)
CTR	Control experiment	920.6	53.3	55.5
E45M	ΔSST changes from -2 to 0°C linearly from eye to r=45 km	922.8	52.8	54.2
E90M	ΔSST changes from -2 to 0°C linearly from eye to r=90 km	940.8	46.1	77.8
E180M	ΔSST changes from -2 to 0°C linearly from eye to r=180 km	951.0	39.4	76.7
E270M	ΔSST changes from -2 to 0°C linearly from eye to r=270 km	952.6	40.4	60.3
EallM	ΔSST=-2°C for all the simulation domain	941.3	43.8	47.0
E45P	ΔSST changes from 2 to 0°C linearly from eye to r=45 km	917.4	53.1	47.5
E90P	ΔSST changes from 2 to 0°C linearly from eye to r=90 km	896.3	60.3	39.0
E180P	ΔSST changes from 2 to 0°C linearly from eye to r=180 km	897.8	60.1	45.3
E270P	ΔSST changes from 2 to 0°C linearly from eye to r=270 km	921.3	55.5	65.1
EallP	ΔSST =2°C for all the simulation domain	915.4	50.2	93.7

10.4 An explanation of relative and absolute SST on TC intensity

10.4.1 *Response of SEF related to SST*

Vecchi *et al.* [2008] and Ramsay and Sobel [2011] demonstrated that it is not the absolute SST but the relative SST that plays a critical role in determining TC intensity. However, the role of SST and the possible physical processes at work on changing storm intensity are still widely debated, though the "bridge effect" of the tropospheric temperature response to relative SST in determining the PI has been taken into consideration [Vecchi and Soden, 2007; Ramsay and Sobel, 2011].

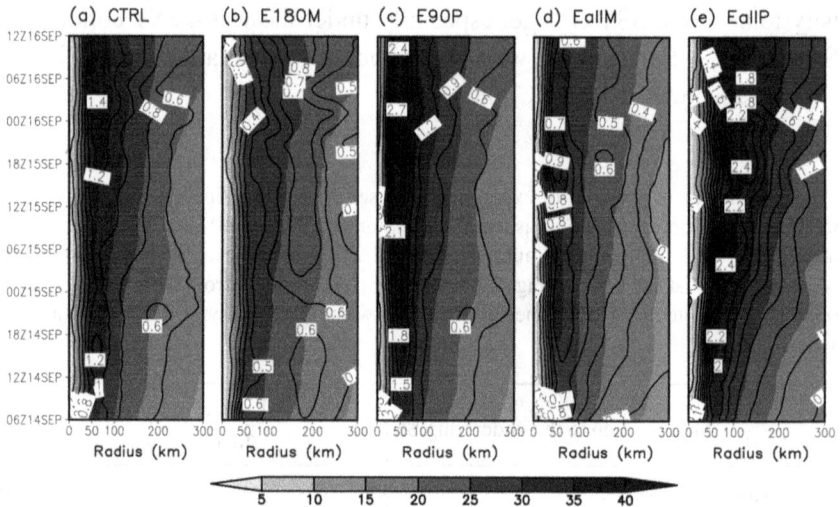

Fig. 10.1. Radius–time cross section of the azimuthal mean surface tangential wind speed (m s^{-1}; shadings) and SEFs (10^3W m^{-2}; contours) in given experiments (a: CTR, b: E180M, c: E90P, d: EallM, and e: EallP).

Since SEF and thus TC intensity is directly related to SST, we first examine how SEF responds to the modification of SST in different experiments. Fig. 10.1 illustrates the radius–time cross section of the azimuthal mean surface tangential wind speed and SEF in some given experiments. It clearly shows that the azimuthal mean SEF in EallP (Fig. 10.1e) is much smaller than that in E90P (Fig. 10.1c) near the eyewall region and EallP gives a much weaker azimuthal mean surface tangential wind speed, and thus a much weaker storm. On the other hand, although the magnitude of the azimuthal mean SEF in EallM is similar to that in E180M, the former is mainly concentrated in the vicinity of the eyewall (Fig. 10.1d), which can help to sustain storm intensity by inputting energy to the eyewall directly, whereas the SEF in E180M decentralizes in a large area outside the eyewall (Fig.

10.1b), and thus contributes little to TC intensity [Xu and Wang, 2010a]. Meanwhile, through the outer tangential winds in E180M and EallP are notably stronger than those in EallM and E90P, the TC intensities in formers is much weaker than those in latters, which is consistent with that suggested by Xu and Wang [2010b] indicating that outer strong wind fields can favor the activity of outer spiral rainbands and thus reduce TC intensity.

10.4.2 *Air–sea temperature and moisture differences*

In order to give an explanation as to why the SEF responds to the absolute SST and relative SST in a different manner, we examine the impacts of SST on sensible-heat flux and latent-heat flux, respectively. Besides, SST can directly affect the air–sea temperature difference (ASTD), which is defined as SST minus air temperature at 2 m (T_2), and thus the sensible-heat flux; it can influence the saturation specific humidity, and thus the air–sea moisture difference (ASMD), which is defined as the saturation specific humidity minus air specific humidity at 2 m (M_2); and determines the latent-heat flux. This is also the key factor in understanding the reason that TC intensity varies in the sensitivity experiments.

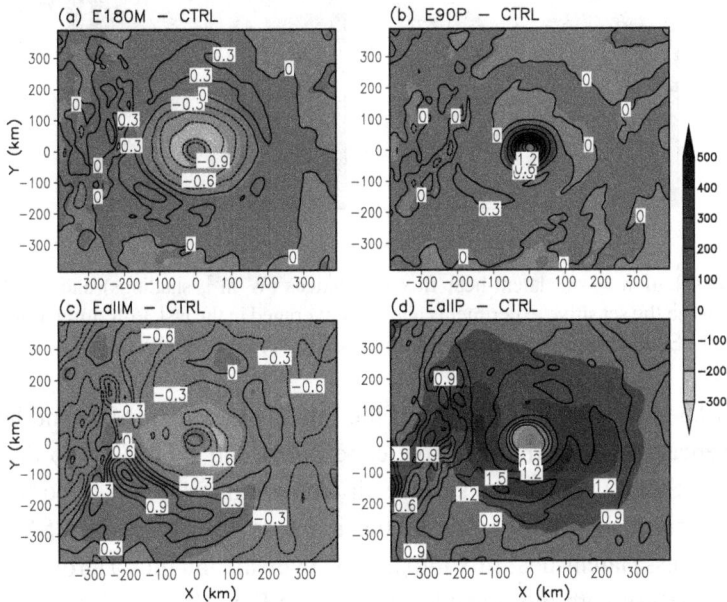

Fig. 10.2. Differences in spatial distribution of sensible-heat flux (W m^{-2}, shadings) and ASTD (°C; contours) between the given sensitive experiments and CTR averaged in the last 36-h simulation. (a) E180M minus CTR; (b) E90P minus CTR; (c) EallM minus CTR; (d) EallP minus CTR.

Fig. 10.2 shows the distribution of the time-averaged ASTDs and sensible-heat flux differences between the sensitive experiment and CTR. Due to the decrease of SST, the decreased ASTD leads to less sensible-heat flux entering the TC eyewall from the ocean and thus a weaker final TC (Figs. 10.2a and 10.2c), while for the increase of SST, the increased ASTD leads to more energy entering the TC eyewall and thus a stronger TC (Figs. 10.2b and 10.2d). However, as shown in Fig. 10.2, with the same magnitude of SST change, the ASTD and sensible-heat flux in the relative SST experiments are notably greater than those in the absolute SST experiments.

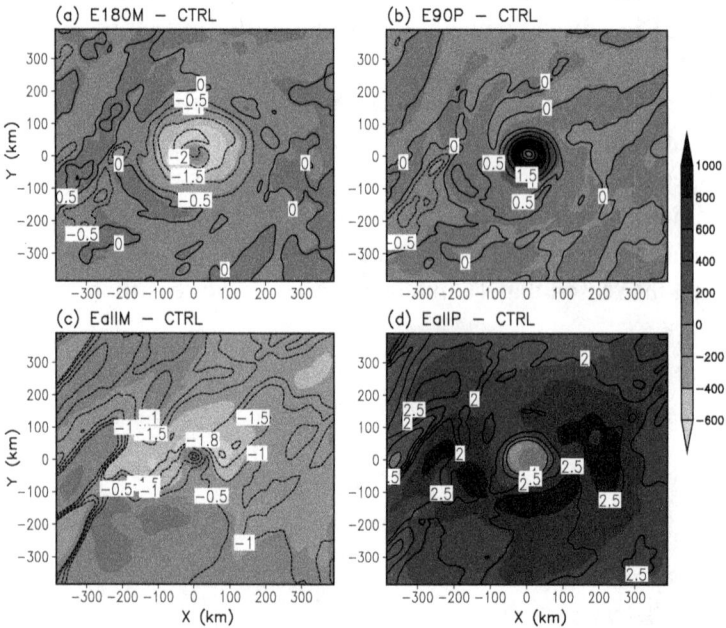

Fig. 10.3. Distribution of the latent-heat flux differences (W m^{-2}, shadings) and ASMD (g kg^{-1}, contours) between the sensitive experiments and CTR averaged in the last 36-h simulation.(a) E180M minus CTR; (b) E90P minus CTR; (c) EallM minus CTR; (d) EallP minus CTR.

Fig. 10.3 shows the distribution of the time-averaged ASMDs and latent-heat flux differences between the sensitive experiment and CTR. Due to the increase (decrease) of SST as well as the surface air temperature, the increased (decreased) saturation specific humidity leads to the increase (decrease) of ASMD, which brings more (less) latent-heat flux into the TC eyewall and thus a stronger (weaker) TC (Fig. 10.3). As with the sensible-heat flux in Fig. 10.2, the latent-heat flux in the relative SST experiments is notably greater than that in the absolute SST experiments.

In order to investigate the differences in SEFs between the relative SST experiments and the absolute SST experiments (Figs. 10.2 and 10.3), the role of remote SST outside CR on TC intensity and the possible involved dynamic and thermodynamic processes are discussed by comparing the relative SST experiments with the absolute SST experiments. In E90P, due to the unchanged SST outside CR, it clearly shows that T_2 and M_2 outside CR are also unchanged and remain low as those in CTR (Figs. are not shown), resulting in unchanged ASTDs and ASMDs in the areas (Figs. 10.2b and 10.3b). When the relatively colder and drier boundary-layer inflow enters the eyewall where the warmer local SST inside CR is located, the increased ASTDs and ASMDs will lead to more energy (i.e., SEF) entering the TC eyewall, which would intensify the TC even more. In EallP, however, for the increase of the absolute SST in the whole simulation region, the ASTMDs are notable outside CR (Figs. 10.2d and 10.3d). As the inflow air gets closer to the eyewall, surface T_2 and M_2 increase gradually for the heating of underlying warmer remote SST, which decreases the ASTMDs as well as the SEF. Thereby, near the eyewall, the ASTDs and ASMDs are much smaller than those in E90P, thus leading to less energy input (i.e., SEF) in the TC eyewall, resulting in a much weaker storm than that in E90P.

On the other hand, comparing the colder relative SST experiment (E180M) with the colder absolute SST experiment (EallM), it can be seen that, though the ASTMDs decrease more in E180M (Figs. 10.2a and 10.3a), the difference in storm intensity between these two experiments are notably smaller than that between E90P and EallP after some initial adjustments (Table 10.1). It suggests that the TC heat potential, a measure of upper-ocean heat content from the surface to the 26°C isotherm depth [Leipper and Volgenau, 1972], is expected to decrease drastically, and thus the remote SST may play a less important role in determining TC intensity. When the SST decreases by 2°C (from 29°C to 27°C), less sensitivity of the TC intensity to the remote SST outside CR appears and the decrease of the remote SST no longer induces a notable change in TC intensity (Table 10.1).

According to the effect of the relative SST and absolute SST on TC intensity, the schematic diagrams for the possible dynamic and thermodynamic processes can be concluded as Fig. 10.4. It suggests that the warmer (colder) relative SST makes the ASTDs and ASMDs, as well as the SEFs, near the eyewall increase (decrease) significantly, which will lead to a notable increase (decrease) in TC intensity (Figs. 10.4a and 10.4b). Whereas the warmer (colder) absolute SST makes the ASTDs, ASMDs and SEFs increase (decrease) mainly in the outer region, which induces relatively less (more) fluxes entering the eyewall, and thus prevents the TC from further intensification (weakening) (Figs. 10.4c and 10.4d), nevertheless, comparing the relative SST experiments with the absolute SST experiments, it can be found that, as inflow air gets closer to the eyewall, the warmer (colder) remote SST outside CR could increase (decrease) the surface air temperature and humidity gradually, and thus decrease (increase) the ASTMDs. This will lead to less (more) fluxes entering the eyewall, as suggested by Emanuel [1986], it will decrease (increase) TC intensity.

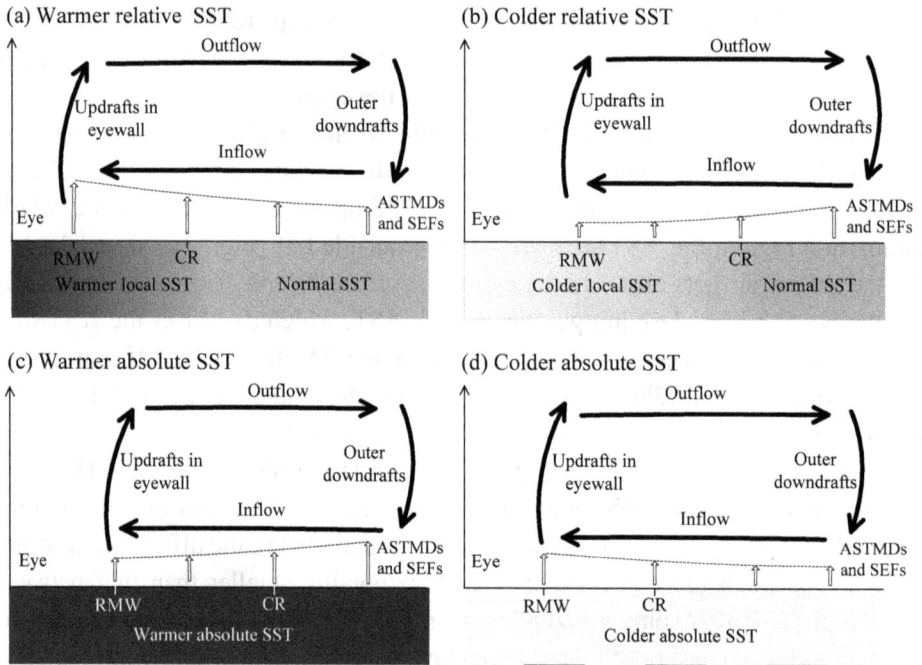

Fig. 10.4. Schematic diagrams showing the thermodynamic processes responsible for (a) the sharp increase in intensity of the simulated TC in the warmer relative SST experiment (such as E90P), (b) the sharp decrease in TC intensity in the colder relative SST experiment (such as E180M), (c) the slight increase in TC intensity in the warmer absolute SST experiment (such as EallP), and (d) the significant decrease in TC intensity in the colder absolute SST experiment (such as EallM).

10.5 Concluding remarks

With sensitivity experiments of relative SST and absolute SST on TC, the role of SST on TC intensity and the possible involved dynamic and thermodynamic processes are analysed. It is found that the relative SST inside CR, a region within a radius of 2–3 times the RMW from the TC center, contributes greatly to TC intensity, but it limits the inner-core size of the TC. In addition, the remote SST outside CR plays a crucial role in the reduction of storm intensity, which will lead to a reduced sensitivity in TC intensity to absolute SST change, especially under warm SST conditions. Moreover, the dependence of storm intensity on the change of SST can be illustrated based on the ASTDs and ASMDs. Due to the decrease (increase) in underlying SST, the decreased (increased) ASTDs and ASMDs lead to less (more) energy (i.e. SEFs) entering the TC eyewall from the ocean and thus a weaker (stronger) final TC.

Here, the possible reason that TC intensity is more sensitive to relative SST inside CR than to absolute SST is proposed. As the inflow air gets closer to the eyewall, the warmer (colder) remote SST outside CR can gradually increase (decrease) the surface air temperature and moisture, and thus decrease (increase) the ASTDs and ASMDs, which will lead to less (more) energy entering the eyewall and decrease (increase) the TC intensity and make TC intensity less sensitive to absolute SST change.

Bibliography

Anthes, R. A. (1977). A cumulus parameterization scheme utilizing a one-dimensional cloud model, *Mon. Wea. Rev.*, 117, pp. 1423–1438.

Anthes, R. A., Hsie, Y. and Kuo, Y. H. (1987) *Description of the Penn State/NCAR mesoscale model version4 (MM4)*. NCAR technical note, NCAT/TN-282+STR, Boulder, Colorado, 66pp.

Arakawa, A., Jung, J.-H. and Wu, C.-M. (2011). Toward unification of the multiscale modeling of the atmosphere, *Atmos. Chem. Phys.*, 11, pp. 3731–3742.

Arakawa, A. and Wu, C.-M. (2013). A unified representation of deep moist convection in numerical modeling of the atmosphere. Part I: *J. Atmos. Sci.*, 70, pp. 1977–1992.

Bell, M. M. and Montgomery. M. T. (2008). Observed structure, evolution, and potential intensity of category 5 Hurricane Isabel from 12 to 14 September, *Mon. Wea. Rev.*, 136, pp. 2023–2046.

Beljaars, A. C. M. (1994). The parameterization of surface fluxes in large-scale models under free convection, *Quart. J. Roy. Meteorol. Soc.*, 121, pp. 255–270.

Bender, M. A., Tuleya, R. E. and Kurihara Y. (1987). A numerical study of the effect of an island terrain on tropical cyclones, *Mon. Wea. Rev.*, 115, pp. 130–155.

Betts, A. K. (1986). A new convective adjustment scheme. Part I: Observational and theoretical basis, *Quart. J. R. Meteor. Soc.*, 112, pp. 677–691.

Betts, A. K. and Miller, M. J. (1986). A new convective adjustment scheme. Part II: Single column tests using GATE wave, BOMEX, ATEX and arctic air-mass data sets, *Quart. J. R. Meteor. Soc.*, 112, pp. 693–709.

Bi, M., Li, T., Peng, M. and Shen, X. (2015). Interactions between Typhoon Megi (2010) and a low-frequency monsoon gyre, *J. Atmos. Sci.*, 72, pp. 2682–2702.

Bister, M. and Emanuel, K. A. (1998). Dissipative heating and hurricane intensity, *Meteor. Atmos. Phys.*, 65, pp. 233–240.

Black, P. G. and Holland, G. J. (1995). The boundary layer of Tropical Cyclone Kerry (1979), *Mon. Wea. Rev.*, 123, pp. 2007–2028.

Brand, S. (1970). Interaction of binary tropical cyclones of the western North Pacific ocean, *J. Appl. Meteorol., 9,* pp. 433–441.

Braun S. A. and Tao, W. K. (2000). Sensitivity of high-resolution simulations of hurricane Bob (1991) to planetary boundary layer parameterizations, *Mon. Wea. Rev.*, 128, pp. 3941–3961.

Braun, S. A. (2006). High-resolution simulation of Hurricane Bonnie (1998). Part II: Water budget, *J. Atmos. Sci.*, 63, pp. 43–64.

Brennan, M. J., and Majumdar, S. J. (2011). An examination of model track forecast errors for Hurricane Ike (2008) in the Gulf of Mexico, *Wea. Forecasting*, 26, pp. 848–867.

Bright, D. R. and Mullen, S. L. (2002). The sensitivity of the numerical simulation of the southwest monsoon boundary layer to the choice of PBL turbulence parameterization in MM5, *Wea. and Forecasting.*, 17, pp. 99–114.

Brown, A. R. (1996). Evaluation of parametrization schemes for the convective boundary layer using large-eddy simulation results, *Bound.-Layer Meteorol.*, 81, pp. 167–200.

Bryan, G. H., Wyngaard, J. C. and Fritsch, J. M. (2003). Resolution requirements for the simulation of deep moist convection, *Mon. Wea. Rev.*, 131, pp. 2394–2416.

Bu, Y. P., Fovell, R. G. and Corbosiero, K. L. (2014). Influence of Cloud–Radiative Forcing on Tropical Cyclone Structure, *J. Atmos. Sci.*, 71, pp. 1644–1662.

Carr, L. E. III and Elsberry, R. L. (1990). Observational evidence for predictions of tropical cyclone propagation relative to steering, *J. Atmos. Sci.*, 47, pp. 542–546.

Carr, L. E. III and Elsberry, R. L. (1997). Models of tropical cyclone wind distribution and beta-effect propagation for application to tropical cyclone track forecasting, *Mon. Wea. Rev.*,125, pp. 3190–3209.

Carr L. E. III and Elsberry, R. L. (2000). Dynamical tropical cyclone track forecast errors. Part II: Midlatitude circulation influences, *Wea. Forecasting*, 15, pp. 662–681.

Carr, L. E. III and Peak, J. E. (2001). Beta test of the systematic approach expert system prototype as a tropical cyclone track forecasting aid, *Wea. Forecasting*, 16, pp. 355–368.

Cha, D.-H., Jin, C.-S., Lee, D.-K. and Kuo, Y.-H. (2011). Impact of intermittent spectral nudging on regional climate simulation using Weather Research and Forecasting model, *J. Geophys. Res.*, 116, D10103.

Chan, J. C. L. (1984). An observational study of the physical processes responsible for tropical cyclone motion, *J. Atmos. Sci.*, 41, pp. 1036–1048.

Chan, J. C. L., Duan, Y. and Shay, L. K. (2001). Tropical cyclone intensity change from a simple ocean–atmosphere coupled model, *J. Atmos. Sci.*, 58, pp. 154–172.

Chan, J. C. L. and Gray, W. M. (1982). Tropical cyclone movement and surrounding flow relationships, *Mon. Wea. Rev.*, 110, pp. 1354–1374.

Chang, S. W.-J. (1982). The orographic effects induced by an island mountain range on propagating tropical cyclones, *Mon. Wea. Rev.*, 110, pp. 1255–1270.

Chen, S.-H. and Sun, W.-Y. (2002). A one-dimensional time dependent cloud model, *J. Meteorol. Soc. Japan*, 80, pp. 99–118.

Chen, S., Campbell, T. J., Jin, H., Gaberšek, S., Hodur, R. M. and Martin, P. (2010). Effect of two-way air–sea coupling in high and low wind speed regimes, *Mon. Wea. Rev.*, 138, pp. 3579–3602.

Christopher, A. D. and Simon, Low-Nam (2001) *The NCAR-AFWA tropical cyclone bogussing scheme*, A Report Prepared for the Air Force Weather Agency (AFWA). National Center for Atmospheric Research, Boulder, Colorado, 13pp.

Chou M.-D. and Suarez, M. J. (1994). An efficient thermal infrared radiation parameterization for use in general circulation models. *NASA Tech. Memo.* 104606, 3, 85 pp.

Craig, G. C. and Dörnbrack, A. (2008). Entrainment in cumulus clouds: What resolution is cloud-resolving?, *J. Atmos. Sci.*, 65, pp. 3978–3988.

Davis, C., Wang, W., Chen, S. S., Chen, Y. and Corbosiero, K. (2008). Prediction of landfalling hurricanes with the Advanced Hurricane WRF model, *Mon. Wea. Rev.*, 136, pp. 1990–2005.

Davies H. C., and Turner, R. E. (1977). Updating prediction methods by dynamical relaxation: An examination of the technique, *Quart. J. Roy. Meteor. Soc.*, 103, pp. 225–245.

Deng, A. and Stauffer, D. R. (2006). On improving 4-km mesoscale model simulations, *J. Appl. Meteor. Climatol.*, 45, pp. 361–381.

Deng, G., Zhou, Y. S. and Li, J. T. (2005). The experiments of the boundary layer schemes on simulated typhoon. Part I: The effect on the structure of typhoon, *Chin. J. of Atmos. Sci.*, 29, pp. 813–824.

Denis, B., Laprise, R., Caya, D. and Côté, J. (2002). Downscaling ability of one-way nested regional climate models: the Big-Brother Experiment, *Clim. Dyn.*, 18, pp. 627–646.

Diaconescu, P. E., Laprise, R. and Sushama, L. (2007). The impact of lateral boundary data errors on the simulated climate of a nested regional climate model, *Clim. Dyn.*, 28, pp. 333–350.

Dickinson, R. E., Errico, R. M., Giorgi, F. and Bates, G. T. (1989). A regional climate model for the western United States, *Climatic Change*, 15, pp. 383–422.

Dickinson R. E., Sellers, A. H. and Kennedy P. J. (1993). *Biosphere-Atmosphere Transfer Scheme (BATS) version 1 as coupled to the NCAR Community Climate Model.* NCAR Tech. Note NCAR/TN-387+STR, Boulder, Colorado, 72 pp.

Ding, Y. H., Shi, X. L., Liu, Y. M., Liu, Y., Li, Q. Q., Qian, Y. F., Miao, M. Q., Zhai, G. Q. and Gao, K. (2006). Multi-year simulations and experimental seasonal predictions for rainy seasons in China by using a nested regional climate model (RegCM_NCC). Part I: Sensitivity study, *Adv. Atmos. Sci.*, 23, pp. 323–341.

Dudhia, J. (1989). Numerical study of convection observed during the winter monsoon experiment using a mesoscale two-dimensional model, *J. Atmos. Sci.*, 46, pp. 3077–3107.

Dyer, A. J. and Hicks, B. B. (1970). Flux-gradient relationships in the constant flux layer, *Quart. J. Roy.* Meteorol. *Soc.*, 96, pp. 715–721.

Eliassen, A. (1951). Slow thermally or frictionally controlled meridional circulation in a circular vortex, *Astrophys. Norv.*, 5, pp. 19–60.

Emanuel, K. A. (1986). An air–sea interaction theory for tropical cyclones. Part I: Steady-state maintenance, *J. Atmos. Sci.*, 43, pp. 585–604.

Emanuel, K. A. (1991a). A scheme for representing cumulus convection in large-scale models, *Quart. J. Roy. Meteor. Soc.*, 48, pp. 2313–2335.

Emanuel, K. A. (1991b). The theory of hurricanes, *Ann. Rev. Fluid Mech.*, 23, pp. 179–196.

Emanuel, K. A. (1995). Sensitivity of tropical cyclones to surface exchange coefficients and a revised steady-state model incorporating eye dynamics, *J. Atmos. Sci.*, 52, pp. 3969–3976.

Fierro, A. O., Rogers, R. F. and Marks, F. D. (2009). The impact of horizontal grid spacing on the microphysical and kinematic structures of strong tropical cyclones simulated with the WRF-ARW model, *Mon. Wea. Rev.*, 137, pp. 3717–3743.

Fiorino, M. and Elsberry, R. L. (1989). Some aspects of vortex structure in tropical cyclone motion, *J. Atmos. Sci.*, *46*, pp. 979–990.

Fovell, R. G. and Su, H. (2007). Impact of cloud microphysics on hurricane track forecasts, *Geophys. Res. Lett.*, *34*, L24810.

Fovell, R. G., Corbosiero, K. L. and Kuo, H.-C. (2009). Cloud microphysics impact on hurricane track as revealed in idealized experiments, *J. Atmos. Sci.*, 66, pp. 1764–1778.

Fudeyasu, H., Wang, Y., Satoh, M., Nasuno, T., Miura, H. and Yanase, W. (2010). Multi-scale interactions in the life cycle of a tropical cyclone simulated in a global cloud-system resolving model. Part I: Large-scale and storm-scale evolutions, *Mon. Wea. Rev.*, 138, pp. 4285–4304.

Galarneau JR, T. J. and Davis, C. A. (2013). Diagnosing forecast errors in tropical cyclone motion, *Mon. Wea. Rev.*, 141, pp. 405–430.

Gentry, M. S. and Lackmann, G. M. (2010). Sensitivity of simulated tropical cyclone structure and intensity to horizontal resolution, *Mon. Wea. Rev.*, 138, pp. 688–704.

Gerard, L. (2007). An integrated package for subgrid convection, clouds and precipitation compatible with meso-gamma scales, *Quart. J. Roy. Meteorol. Soc.*, 133, pp. 711–730.

Giorgi, F. (1990). Simulation of regional climate using a limited area model nested in general circulation model, *J. Climate*, 3, pp. 941–963.

Giorgi, F., and Mearns, L.O. (1991). Approaches to the simulation of regional climate change: a review, *Rev. Geophys.*, 29, pp. 191–216.

Giorgi, F. and Marinucci, M. R. (1996). An investigation of the sensitivity of simulated precipitation to model resolution and its implications for climate studies, *Mon. Wea. Rev.*, 124, pp. 148–166.

Giorgi, F., Huang, Y., Nishizawa, K. and Fu, C. (1999). A seasonal cycle simulation over eastern Asia and its sensitivity to radiative transfer and surface processes, *J. Geophys. Res.*, 104, pp. 6403–6423.

Giorgi, F., Marinucci, M. R. and Bates, G. T. (1993a). Development of a second-generation regional climate model (RegCM2). Part I: Boundary-layer and radiative transfer processes, *Mon. Wea. Rev.*, 121, pp. 2794–2813.

Giorgi, F., Marinucci, M. R., Bates, G. T. and De Canio, G. (1993b). Development of a second-generation regional climate model (RegCM2). Part II: Convective processes and assimilation of lateral boundary conditions, *Mon. Wea. Rev.*, 121, pp. 2814–2832.

Giorgi, F. and Mearns, L.O. (1999). Introduction to special section: Regional climate modeling revisited, *J. Geophys. Res.*, 104, pp. 6335–6352.

Gong, W. and Wang, W.-C. (2000). A regional model simulation of the 1991 severe precipitation event over the Yangtze-Huai valley. Part II: Model bias, *J. Clim.*, 13, pp. 93–108.

Gopalakrishnan, S. G., Marks, F. Jr., Zhang, X. J., Bao, J.-W., Yeh, K.-S. and Atlas, R. (2011). The experimental HWRF system: A study on the influence of horizontal resolution on the structure and intensity changes in tropical cyclones using an idealized framework, *Mon. Wea. Rev.*, 139, pp. 1762–1784.

Grell, G. A. (1993). Prognostic evaluation of assumptions used by cumulus parameterizations, *Mon. Wea. Rev.*, 121, pp. 764–787.

Grell, G. A. and Dévényi, D. (2002). A generalized approach to parameterizing convection combining ensemble and data assimilation techniques, *Geophys. Res. Lett.*, 29, pp. 381–384.

Grell, G. A., Dudhia, J. and Stauffer, D. R. (1994). *A description of the fifth generation Penn State/NCAR Mesoscale Model (MM5)*. NCAR Tech. Note NCAR/TN 398+STR, Boulder, Colorado, 121 pp.

Grell, G. A. and Freitas, S. (2013). A scale and aerosol aware stochastic convective parameterization for weather and air quality modeling, *Atmos. Chem. Phys. Discuss.*, 13, pp. 23846–23893.

Grell, E., Grell, G. and Bao, J.-W. (2013). Experimenting with a convective parameterization scheme suitable for high-resolution mesoscale models in tropical cyclone simulations, *Geophys. Res. Abs.*, 15, EGU2013-5746-2.

Guan, H., Wang, H. J., Zhou, L., et al. (2011). A numerical simulation study on the typhoon-ocean interaction in the South China Sea, *Chin. J. Geophys.*, 54, pp. 1141–1149.

Guinn, T. A. and Schubert, W. H. (1993). Hurricane spiral bands. *J. Atmos. Sci.*, 50, pp. 3380–3403.

Hack, J. J. and Schubert, W. H. (1986). Nonlinear response of atmospheric vortices to heating by organized cumulus convection, *J. Atmos. Sci.*, 43, pp. 1559–1573.

Hammarstrand, U. (1998). Questions involving the use of traditional convection parameterization in NWP models with higher resolution, *Tellus*, 50, pp. 265–282.

Han, Y. and Wu, R. S. (2008). The effect of cold air intrusion on the development of tropical cyclone, *Chin. J. Geophys.*, 51, pp. 321–1332.

Hazelton, A. T. and Hart, R. E. (2013). Hurricane eyewall slope as determined from airborne radar reflectivity data: Composites and case studies, *Wea. Forecasting*, 28, pp. 368–386.

Hendricks, E. A., Montgomery, M. T. and Davis, C. A. (2004). The role of "vortical" hot towers in the formation of tropical cyclone Diana (1984), *J. Atmos. Sci.*, 61, pp. 1209–1232.

Hill, K. A. and Lackmann, G. M. (2009). Influence of Environmental Humidity on Tropical Cyclone Size, *Mon. Wea. Rev.*, 137, pp. 3294–3315.

Hogan, T. F. and Pauley, R. L. (2007). The impact of convective momentum transport on tropical cyclone track forecasts using the Emanuel cumulus parameterization, *Mon. Wea. Rev.*, 135, pp. 1195–1207.

Holland, G. J. (1983). Tropical cyclone motion: Environmental interaction plus a beta effect, *J. Atmos. Sci.*, 40, pp. 328–342.

Holland, G. J. (1993). *Tropical cyclone motion. Global Guide to Tropical Cyclone Forecasting*, World Meteorological Organization Tech. Document WMO/TD 560, Tropical Cyclone Programme Rep. TCP-31, Geneva, Switzerland. [Available online at http://www.bom.gov.au/bmrc/pubs/tcguide/globapguidepintro.htm]

Holland, G. J. (1997). The maximum potential intensity of tropical cyclones, *J. Atmos. Sci.*, 54, pp. 2519–2541.

Holland, G. J. and Merrill, R. T. (1984). On the dynamics of tropical cyclone structural changes, *Quart. J. Roy. Meteor. Soc.*, 40, pp. 723–745.

Holtslag, A. A. M., De Bruijin, E. I. F. and Pan, H. L. (1990). A high resolution air mass transformation model for short-range weather forecasting, *Mon. Wea. Rev.*, 118, pp. 1561–1575.

Hong, S.-Y. and Choi, J. (2006). Sensitivity of the simulated regional climate circulations over east Asia in 1997 and 1998 summers to three convective parameterization schemes, *J. Kor. Meteor. Soc.*, 42, pp. 361–378.

Hong, S.-Y., Dudhia, J. and Chen, S.-H. (2004). A revised approach to ice microphysical processes for the bulk parameterization of clouds and precipitation, *Mon. Wea. Rev.*, 132, pp. 103–120.

Hong, S.- Y., Juang, H. M. H. and Lee, D.-K. (1999). Evaluation of a regional spectral model for the East Asia monsoon case studies for July 1987 and 1988, *J. Meteor. Soc. Japan*, 77, pp. 553–572.

Hong, S. Y. and Lim, J.-O. J. (2006). The WRF single-moment 6-class microphysics scheme (WSM6), *J. Korean Meteor. Soc.*, 42, pp. 129–151.

Hong, S. Y., Noh, Y. and Dudhia, J. (2006). A new vertical diffusion package with an explicit treatment of entrainment processes, *Mon. Wea. Rev.*, 134, pp. 2318–2341.

Hoskins, B. J. (1996). On the existence and strength of the summer subtropical anticyclones, *Bull. Amer. Meteor. Soc.*, 77, pp. 1287–1292.

Houtekamer, P. L., Lefaivre, L., Derome, J., Ritchie, H. and Mitchell, H. L. (1996). A system simulation approach to ensemble prediction, *Mon. Wea. Rev.*, 124, pp. 1225–1242.

Houze, R. A. Jr., Chen, S. S. and Lee, W. C. (2006). The hurricane rainband and intensity change experiment: Observations and modeling of Hurricanes Katrina, Ophelia, and Rita, *Bull. Amer. Meteor. Soc.*, 87, pp. 1503–1521.

Houze, R. A. Jr., Chen, S. S., Smull, B. F. and Bell, M. M. (2007). Hurricane intensity and eyewall replacement, *Science*, 315, pp. 1235–1239.

Hsu, H.-H., Hung, C.-H., Lo, A.-K., Wu, C.-C. and Hung, C.-W. (2008). Influence of tropical cyclones on the estimation of climate variability in the tropical western north Pacific, *J. Clim.*, 21, pp. 2960–2975.

Huang, W. Y., Shen, X. Y., Wang, W. G. (2014). Comparison of the thermal and dynamic structural characteristics in boundary layer with different boundary layer parameterizations, *Chin. J. Geophys.*, 57, pp. 399–1414.

Huffman, G. J., Adler, R. F., Morrissey, M. M., Bolvin, D. T., Curtis, S., Joyce, R., McGavock, B. and Susskind, J. (2001). Global precipitation at one-degree daily resolution from multisatellite observations, *J. Hydrometeor.*, 2, pp. 36–50.

Janjić, Z. I. (1994). The step-mountain eta coordinate model: Further developments of the convection, viscous sublayer and turbulence closure schemes, *Mon. Wea. Rev.*, 122, pp. 927–945.

Janjić, Z. I. (1996). The surface layer in the NCEP Eta Model. Preprints, 11th Conf. on Numerical Weather Prediction, Norfolk, VA, *Amer. Meteor. Soc.*, pp. 354–355.

Janjić, Z. I. (2000). Comments on "Development and evaluation of a convection scheme for use in climate models", *J. Atmos. Sci.*, 57, pp. 1766–1782.

Janjić, Z. I. (2002) Nonsingular implementation of the Mellor-Yamada level 2.5 scheme in the NCEP meso model. *NCEP office Note*, No. 437, 61 pp.

Jin, Y., Wang, S., Nachamkin, J., Doyle, J. D. and Thompson, G. (2014). The impact of ice phase cloud parameterizations on tropical cyclone prediction, Mon. *Wea. Rev.*, 142, pp. 606–625.

Jones, R. G., Murphy, J. M. and Noguer, M. (1995). Simulation of climate change over Europe using a nested regional climate model I: Assessment of control climate, including sensitivity to location of lateral boundaries, *Quart. J. Roy. Meteor. Soc.*, 121, pp. 1413–1449.

Ju, J. and Slingo, J. (1995). The Asian summer monsoon and ENSO, *Quart. J. Roy. Meteor. Soc.*, 121, pp. 1133–1168.

Kuo, H. L. (1965). On formation and intensification of tropical cyclones through latent heat release by cumulus convection, *J. Atmos. Sci.*, 22, pp. 40–63.

Kain, J. S. (2004). The Kain-Fritsch convective parameterization: an update, *J. Appl. Meteorol.*, 43, pp. 170–181.

Kain, J. S. and Fritsch, J. M. (1990). A one-dimensional entraining/detraining plume model and its application in convective parameterization, *J. Atmos. Sci.*, 47, pp. 2784–2802.

Kain, J. S., Weiss, S. J., Bright, D. R., et al. (2008). Some practical considerations for the first generation of operational convection-allowing NWP, *Wea. Forecasting.*, 23, pp. 931–952.

Kalnay, E., Kamanitsu, M., Kistler, R., Cpllins, W., Deaven, D., Gandin, L., Iredell, M., Saha, S., White, G., Woollen, J., Zhu, Y., Leetmaa, A., Reynolds, B., Chelliah, M., Ebisuzaki, W., Higgins, W., Janowiak, J., Mo, K. C., Ropelewski, C., Wang, J., Jenne, R. and Joseph, D. (1996). The NCEP/NCAR 40-year reanalysis project, *Bull. Am. Meteor. Soc.*, 77, pp. 437–472.

Kang, I. S., Jin, K. and Wang, B. (2002). Intercomparison of the climatological variations of Asian summer monsoon precipitation simulated by 10 GCMs, *Clim. Dyn.*, 19, pp. 383–395.

Kato, H., Hirakuchi, H., Nishizawa, K. and Giorgi, F. (1999). Performance of NCAR RegCM in the simulation of June and January climates over eastern Asia and the high-resolution effect of the model, *J. Geophy. Res.*, 104, pp. 6455–6476.

Kiehl, J. T., Hack, J. J., Bonan, G. B., Boville, B. A., Briegleb, B. P., Williamson, D. L. and Rasch, P. J. (1996). *Description of NCAR Community Climate Model (CCM3)*. NCAR Tech. Note NCAR/TN-420+STR, Boulder, Colorado, 152 pp.

Kimball, S. K. (2006). A modeling study of hurricane landfall in a dry environment, *Mon. Wea. Rev.*, 134, pp. 1901–1918.

Komaromi, W. A., Majumdar, S. J. and Rappin, E. D. (2011). Diagnosing initial condition sensitivity of Typhoon Sinlaku (2008) and Hurricane Ike (2008), *Mon. Wea. Rev.*, 139, pp. 3224–3242.

Kotroni, V. and Lagouvardos, K. (2004). Evaluation of MM5 high resolution real-time forecasts over the urban area of Athens, Greece, *J. Appl. Meteorol.*, 43, pp. 1666–1678.

Kubota, H., Shirooka, R., Ushiyama, T., Chuda, T., Iwasaki, S. and Takeuchi, K. (2005). Seasonal variations of precipitation properties associated with the monsoon over Palau in the western Pacific, *J. Hydrometeor.*, 6, pp. 518–531.

Kubota, H. and Wang, B. (2009). How much do tropical cyclones affect seasonal and interannual rainfall variability over the western North Pacific?, *J. Clim.* 22, pp. 5495–5510.

Lai, W. F., Liu, Y., Mai, J. H. (2010). Sensitivity numerical simulation of typhoon "Molave" with different boundary layer parameterizations, *Guangdong Meteorology.*, 32, pp. 10–14. (in Chinese)

Lau, K. M. and Yang, S. (1996). Seasonal variation, abrupt transition, and intraseasonal variability associated with the Asian monsoon in the GLA GCM, *J. Clim.*, 9, pp. 965–985.

Lean, H. W., Clark, P. A., Dixon, M., Roberts, N. M., Fitch, A., Forbes, R. and Halliwell, C. (2008). Characteristics of high-resolution versions of the Met Office Unified Model for forecasting convection over the United Kingdom, *Mon. Wea. Rev.*, 136, pp. 3408–3424.

Lee, C.-S., Cheung, K. K. W., Fang, W.-T. and Elsberry, R. L. (2010). Initial maintenance of tropical cyclone size in the western North Pacific, *Mon. Wea. Rev.*, *138*, pp. 3207–3223.

Lee, C. Y. and Chen, S. S. (2012). Symmetric and asymmetric structures of hurricane boundary layer in coupled atmosphere–wave–ocean models and observations, *J. Atmos. Sci.*, 69, pp. 3576–3594.

Lee, D.-K. and Suh, M.-S. (2000). The ten-year east Asian summer monsoon simulation using a regional climate model (RegCM2), *J. Geophys. Res.*, 105, pp. 29565–29577.

Lee, D.-K., Kang, H.-S. and Min, K.-H. (2002). The role of ocean roughness in regional climate modeling: 1994 East Asia summer monsoon case, *J. Meteor. Soc. Japan*, 80, pp. 171–189.

Lee, D.-K., Ahn, Y.-I. and Kim, C.-J. (2004). Impact of ocean rough and bogus typhoons on summertime circulation in a wave-atmosphere coupled regional climate model, *J. Geophys. Res.*, 109, pp. 539–547.

Leipper, D. F. and Volgenau, D. (1972). Hurricane heat potential of the Gulf of Mexico, *J. Phys. Oceanogr.*, 2, pp. 218–224.

Leung, L. R., Ghan, S. J., Zhao, Z.-C., Luo, Y., Wang, W.-C. and Wei, H.-L. (1999). Intercomparison of regional climate simulations of the 1991 summer monsoon in eastern Asia, *J. Geophys. Res.*, 104, pp. 6425–6454.

Leung, L. R., Mearns, L. O., Giorgi, F., and Wilby, R. L. (2003). Regional climate research needs and opportunities, *Bull. Am. Meteor. Soc.*, 84, pp. 89–95.

Leung, L. R., Zhong, S. Y., Qian, Y. and Liu, Y. M. (2004). Evaluation of regional climate simulations of 1998 and 1999 East Asian summer monsoon using

GAME/HUBEX observational data, *J. Meteor. Soc. Japan*, 82, pp. 1695–1713.

Lester, E. C. III and Elsberry, R. L. (1997). Models of tropical cyclone wind distribution and beta-effect propagation for application to tropical cyclone track forecasting, *Mon. Wea. Rev.*, 125, pp. 3190–3209.

Lester E. C. III and Elsberry, R. L. (2000). Dynamical tropical cyclone track forecast errors. Part I: Tropical region error sources, *Wea. Forecasting*, 15, pp. 641–661.

Li, C. H., Huang, F. J. and Luo, Z. X. (2002). The influence of typhoon on subtropical high location and intensity, *Plateau Meteorology.*, 21, pp. 576–582. (in Chinese)

Li, Y. and Chen, L. S. (2005). Numerical study on impacts of boundary layer fluxes over wetland on sustention and rainfall of landfalling tropical cyclone, *Acta Meteorologica Sinica*, 63, pp. 683–693.

Liang, X.-Z., Kunkel, K. E. and Samel, A. N. (2001). Development of a regional climate model for US Midwest applications. Part I: Sensitivity to buffer zone treatment, *J. Clim.*, 14, pp. 4363–4378.

Liang, X.-Z., Xu, M., Yuan, X., *et al.* (2012). Regional climate–weather research and forecasting model, *Bull. Am. Meteor. Soc.*, 93, pp. 1363–1387.

Lin, I.- I., Wu, C.- C., Emanuel, K. A., Lee, I. H., Wu, C. R. and Pun, I. F. (2005). The interaction of Supertyphoon Maemi (2003) with a warm ocean eddy, *Mon. Wea. Rev.*, 133, pp. 2635–2649.

Lin, Y.-L., Ensley, D. B., Chiao, S. and Huang, C.-Y. (2002). Orographic influences on rainfall and track deflection associated with the passage of a tropical cyclone, *Mon. Wea. Rev.*, 130, pp. 2929–2950.

Lin, Y.-L., Farley, R. D. and Orville, H. D. (1983). Bulk parameterization of the snow field in a cloud model, *J. Climate Appl. Meteorol.*, 22, pp. 1065–1092.

Liu, Y., Giorgi, F. and Washington, W. M. (1994). Simulation of summer monsoon climate over East Asia with an NCAR regional climate model, *Mon. Wea. Rev.*, 122, pp. 2331–2348.

Liu, Y., Wu, G., Liu, H., and Liu, P. (2001). Dynamical effects of condensation heating on the subtropical anticyclones in the eastern hemisphere, *Clim. Dyn.*, 17, pp. 327–338.

Liu, Y. B., Zhang, D. L. and Yau, M. K. (1997). A multiscale numerical chapter of Hurricane Andrew (1992). Part I: Explicit simulation and verification, *Mon. Wea. Rev.*, 125, pp. 3073–3093.

Lord, S. J., Willoughby, H. E. and Piotrowicz, J. M. (1984). Role of a parameterized ice-phase microphysics in an axisymmetric, nonhydrostatic tropical cyclone model, *J. Atmos. Sci.*, 41, pp. 2836–2848.

Malkus, J. S. and Riehl, H. (1960). On the dynamics and energy transformations in steady-state hurricane, *Tellus*, 12, pp. 1–20.

Mapes, B. E. and Houze, R. A. (1995). Diabatic divergence profiles in western Pacific mesoscale convective system, *J. Atmos. Sci.*, 52, pp. 1807–1828.

Marbaix, P., Gallée, H., Brasseur, O. and Ypersele, J.-P. (2003). Lateral boundary conditions in regional climate models: a detailed study of the relaxation procedure, *Mon. Wea. Rev.*, 131, pp. 461–479.

Marks, F. D. and Shay, L. K. (1998), Landfalling tropical cyclones: Forecast problems and associated research opportunities, *Bull. Am. Meteorol. Soc.*, 79, pp. 305–323.

Mass, C. F., Ovens, D., Westrick, K. and Colle, B. A. (2002). Does increasing horizontal resolution produce more skillful forecasts?, *Bull. Am. Meteorol. Soc.*, 83, pp. 407–430.

Maclay, K. S., DeMaria, M. and Vonder Haar, T. H. (2008). Tropical cyclone inner-core kinetic energy evolution, *Mon. Wea. Rev.*, 136, pp. 4882–4898.

McCumber, M., Tao, W.-K. and Simpson, J. (1991). Comparison of ice-phase microphysical parameterization schemes using numerical simulations of tropical convection, *J. Appl. Meteorol.*, 30, pp. 985–1004.

McFarquhar, G. M., Zhang, H., Heymsfield, G., Halverson, J. B., Hood, R., Dudhia, J. and Marks, F. (2006). Factors affecting the evolution of Hurricane Erin (2001) and the distributions of hydrometeors: Role of microphysical processes, *J. Atmos. Sci.*, 63, pp. 127–150.

McGregor, J. L. (1997). Regional climate modelling, *Meteor. Atmos. Phys.*, 63, pp. 105–117.

McTaggart-Cowan, R., Bosart, L. F., Gyakum, J. R. and Atallah, E. H. (2006). Hurricane Juan (2003). Part II: Forecasting and numerical simulation, *Mon. Wea. Rev.*, 134, pp. 1748–1771.

Mellor, G. L. and Yamada, T. (1982). Development of a turbulence closure model for geophysical fluid problems, *Rev. Geophys. Space Phys.*, 20, pp. 851–875.

Min, Y., Shen, T. L., Zhu, W. J., et al. (2010). Numerical simulation and diagnosis analysis of spiral rain bands in typhoon "Pearl", *Trans. Atmos. Sci.*, 33, pp. 227–235. (in Chinese)

Miyamoto, Y. and Takemi, T. (2010). An effective radius of the sea surface enthalpy flux for the maintenance of a tropical cyclone, *Atmos. Sci. Lett.*, 11, pp. 278–282.

Mlawer, E. J., Taubman, S. J., Brown, P. D., Iacono, M. J. and Clough, S. A. (1997). Radiative transfer for inhomogeneous atmosphere: RRTM, a validated correlated-k model for the longwave, *J. Geophys. Res.*, 102, pp. 16663–16682.

Molinari, J. M. and Dudek, M. (1992). Parameterization of convective precipitation in mesoscale numerical models: a critical review, *Mon. Wea. Rev.*, 120, pp. 326–344.

Montgomery, M. T., Nicholls, M. E., Cram, T. A. and Saunders, A. (2006). A vertical hot tower route to tropical cyclogenesis, *J. Atmos. Sci.*, 63, pp. 355–386.

Mooney, P. A., Mulligan, F. J. and Fealy, R. (2013). Evaluation of the sensitivity of the Weather Research and Forecasting Model to parameterization schemes for regional climates of Europe over the period 1990–95, *J. Clim.*, 26, pp. 1002–1017.

Murata, A., Saito, K. and Ueno, M. (2003). The effects of precipitation schemes and horizontal resolution on the major rainband in typhoon Flo (1990) predicted by the MRI mesoscale nonhydrostatic model, *Meteor. Atmos. Phys.*, 82, pp. 55–73.

Nakazawa, T. and Rajendran, K. (2007). Relationship between tropospheric circulation over the western North Pacific and tropical cyclone approach/landfall on Japan, *J. Meteor. Soc. Japan*, 85, pp. 101–114.

Ni, Z. P., Lu, W. and Zhang, L. (2013). Analysis on forecasting errors and associated circulations of sudden typhoon track changes during 2005-2010, *Meteorol. Mon.*, 39, pp.719–727. (in Chinese)

Niemelä, S. and Fortelius, C. (2005). Applicability of large-scale convection and condensation parameterization to meso-γ-scale HIRLAM: a case chapter of a convective event, *Mon. Wea. Rev.*, 133, pp. 2422–2435.

Noh, Y., Cheon, W. G., Hong, S. Y., et al. (2003). Improvement of the K-profile Model for the Planetary Boundary Layer based on Large Eddy Simulation Data, *Bound.-Layer Meteorol.*, 107, pp. 401–427.

Nolan, D. S., Zhang, J. A. and Stern, D. P. (2009). Evaluation of Planetary Boundary Layer Parameterizations in Tropical Cyclones by Comparison of In Situ Observations and High-Resolution Simulations of Hurricane Isabel (2003).Part II: Inner-Core Boundary Layer and Eyewall Structure, *Mon. Wea. Rev.*, 137, pp. 3675–3698.

Noone, D. (2012). Pairing measurements of the water vapor isotope ratio with humidity to deduce atmospheric moistening and dehydration in the tropical midtroposphere, *J. Clim.*, 25, pp. 4476–4494.

Noone, D., Galewsky, J., Sharp, Z. D., Worden, J., Barnes, J. (2011). Properties of air mass mixing and humidity in the subtropics from measurements of the D/H isotope ratio of water vapor at the Mauna Loa Observatory, *J. Geophys. Res.*, 116, D22113, pp. 898–908.

Pal, J. S., Giorgi, F., Bi, X., *et al.* (2007). Regional climate modeling for the developing world: The ICTP RegCM3 and RegCNET, *Bull. Am. Meteor. Soc.*, 88, pp. 1385–1409.

Pauley, P. M. and Smith, P. J. (1988). Direct and indirect effects of latent heat release on a synoptic-scale wave system, *Mon. Wea. Rev.*, 116, pp. 1209–1235.

Paulson, C. A. (1970). The mathematical representation of wind speed and temperature profiles in the unstable atmospheric surface layer, *J. Appl. Meteorol.*, 9, pp. 857–861.

Peng, M. S., Jeng, B. F. and Williams, R. T. (1999). A numerical study on tropical cyclone intensification. Part I: Beta effect and mean flow effect, *J. Atmos. Sci.*, 56, pp. 1404–1423.

Persing, J. and Montgomery, M. T. (2005). Is environmental CAPE important in the determination of maximum possible hurricane intensity?, *J. Atmos. Sci.*, 62, pp. 542–550.

Powell, M. D. (1990). Boundary layer structure and dynamics in outer hurricane rainbands. Part I: Mesoscale rainfall and kinematic structure, *Mon. Wea. Rev.*, 118, pp. 891–917.

Powell, M. D. and Reinhold, T. A. (2007). Tropical cyclone destructive potential by integrated kinetic energy, *Bull. Amer. Meteor. Soc.*, 88, pp. 513–526.

Ramsay, H. A. and Sobel, A. H. (2011). Effects of relative and absolute sea surface temperature on tropical cyclone potential intensity using a single-column model, *J. Clim.*, 24, pp. 183–193.

Randall, D. A., Khairoutdinov, M., Arakawa, A. and Grabowski, W. (2003). Breaking the cloud parameterization deadlock, *Bull. Am. Meteor. Soc.*, 84, pp. 1547–1564.

Roberts, N. M. and Lean, H. W. (2008). Scale-selective verification of rainfall accumulations from high-resolution forecasts of convective events, *Mon. Wea. Rev.*, 136, pp. 78–97.

Rogers, R., Aberson, S., Black, M., et al. (2006). The intensity forecasting experiment: A NOAA multiyear field program for improving tropical cyclone intensity forecasts, *Bull. Am. Meteorol. Soc.*, 87, pp. 1523–1537.

Rotunno, R., Chen, Y., Wang, W., Davis, C., Dudhia, J. and Holland, G. J. (2009). Large-eddy simulation of an idealized tropical cyclone, *Bull. Am. Meteor. Soc.*, 90, pp. 1783–1788.

Rotunno, R. and Emanuel, K. A. (1987). An air–sea interaction theory for tropical cyclones. Part II: Evolutionary study using a nonhydrostatic axisymmetric numerical model, *J. Atmos. Sci.*, 44, pp. 542–562.

Rozoff, C. M., Schubert, W. H., McNoldy, B. D. and Kossin, J. P. (2006). Rapid filamentation zones in intense tropical cyclones, *J. Atmos. Sci.*, 63, pp. 435–456.

Rutledge, S. A. and Hobbs, P. V. (1984). The mesoscale and microscale structure and organization of clouds and precipitation in midlatitude cyclones. XII: A diagnostic modeling study of precipitation development in narrow cloud-frontal rainbands, *J. Atmos. Sci.*, *20*, pp. 2949–2972.

Qian, Y. and Leung, L. R. (2007). A long-term regional climate simulation and observations of the hydroclimate in China, *J. Geophys. Res.*, 112, D14104

Samsury, C. E. and Zipser, E. J. (1995). Secondary wind maxima in hurricanes: Airflow and relationship to rainbands, *Mon. Wea. Rev.*, 123, pp. 3502–3517.

Scherrer, S. C., Appenzeller, C., Eckert, P. and Cattani, D. (2004). Analysis of the spread–skill relations using the ECMWF ensemble prediction system over Europe. *Wea. Forecasting*, 19, pp. 552–565.

Seth, A. and Giorgi, F. (1998). The effect of domain choice on summer precipitation simulation and sensitivity in a regional climate model, *J. Clim.*, 11, pp. 2698–2712.

Shay, L. K., Goni, G. J. and Black, P. G. (2000). Effects of a warm oceanic feature on Hurricane Opal, *Mon. Wea. Rev.*, 128, pp. 1366–1383.

Skamarock, W. C., Klemp, J. B., Dudhia, J., Gill, D. O., Barker, D. M., Duda, M. G., Huang, X. Y., Wang, W. and Powers, J. G. (2008). A description of the Advanced Research WRF Version 3, *NCAR Tech.* Note NCAR/TN-475+STR, 113pp. [Available online at http://www.mmm.ucar.edu/people/skamarock]

Sriver, R. L., Huber, M. and Nusbaumer, J. (2008). Investigating tropical cyclone-climate feedbacks using the TRMM Microwave Imager and the Quick Scatterometer, *Geochem. Geophys. Geosyst.*, 9, Q09V11.

Staniforth, A. (1995). Regional modeling: Theoretical discussion, WMO PWPR Rep Ser Vol 7, WMO/TD-No 699 72pp, World Meteorological Organization, Geneva.

Steppeler, J., Doms, G., Schättler, U., Bitzer, H. W., Gassmann, A., Damrath, U. and Gregoric, G. (2003). Meso-gamma scale forecasts using the nonhydrostatic model LM, *Meteorol. Atmos. Phys.*, 82, pp. 75–96.

Stern, D. P. and Nolan, D. S. (2009). Re-examining the vertical structure of tangential winds in tropical cyclones: Observations and theory, *J. Atmos. Sci.*, 66, pp. 3579–3600.

Stossmeister, G. J. and Barnes, G. M. (1992). The development of a second circulation center within tropical storm Isabel (1985), *Mon. Wea. Rev.*, 120, pp. 685–697.

Sun, Y., Yi, L., Zhong, Z., Hu, Y. and Ha, Y. (2013a). Dependence of model convergence on horizontal resolution and convective parameterization in simulations of a tropical cyclone at grey-zone resolutions, *J. Geophys. Res.*, 118, pp. 7715–7732.

Sun, Y., Yi, L., Zhong, Z. and Ha, Y. (2014b). Performance of a new convective parameterization scheme on model convergence in simulations of a tropical cyclone at grey-zone resolutions, *J. Atmos. Sci.*, 71, pp. 2078–2088.

Sun, Y., Zhong, Z., Dong, H., Shi, J. and Hu, Y. (2015a). Sensitivity of tropical cyclone track simulation over the western North Pacific to different heating/drying rates in the Betts–Miller–Janjić scheme, *Mon. Wea. Rev.*, 143, pp. 3478–3494.

Sun, Y., Zhong, Z., Ha, Y., Wang, Y. and Wang, X. D. (2013b). The dynamic and thermodynamic perspectives of relative and absolute sea surface temperature on tropical cyclone intensity, *Acta Meteorol. Sinica*, 27, pp. 40–49.

Sun, Y., Zhong, Z., Lu, W. and Hu, Y. (2014a). Why are tropical cyclone tracks over the western North Pacific sensitive to the cumulus parameterization scheme in regional climate modeling? A case study for Megi (2010), *Mon. Wea. Rev.*, 142, pp. 1240–1249.

Sun, Y., Zhong, Z. and Lu, W. (2015b). Sensitivity of tropical cyclone feedback on the intensity of the western Pacific subtropical high to microphysics schemes, *J. Atmos. Sci.*, 72, pp. 1346–1368.

Sun Y., Zhong, Z. and Wang, Y. (2012). Characteristics of asymmetric flow of tropical cyclone Shanshan (2006) during its turning and intensification period, *Acta Meteorol. Sinica.*, 26, pp. 147–162.

Sun, Y., Zhong, Z. and Yi, L., Li, T. and Chen, M. (2015c). Dependence of the relationship between the tropical cyclone track and western Pacific subtropical high intensity on initial storm size: A numerical investigation, *J. Geophys. Res.-Atmos.*, 120, pp. 11451–11467.

Tao, W.-K., Simpson, J. and McCumber, M. (1989). An ice-water saturation adjustment, *Mon. Wea. Rev.*, 117, pp. 231–235.

Thompson, G., Rasmussen, R. M. and Manning, K. (2004). Explicit forecasts of winter precipitation using an improved bulk microphysics scheme. Part I: Description and sensitivity analysis, *Mon. Wea. Rev.*, 132, pp. 519–542.

Thompson, G., Field, P. R., Rasmussen, R. M. and Hall, W. D. (2008). Explicit forecasts of winter precipitation using an improved bulk microphysics scheme. Part II: Implementation of a new snow parameterization, *Mon. Wea. Rev.*, 136, pp. 5095–5115.

Tao, S. Y. and Chen, L. X. (1987) *Monsoon meteorology*, eds. Chang, C. P. and Krishnamurti, T. N., Chapter 4 "A review of recent research on the East Asian summer monsoon in China", (Oxford University Press, USA) pp. 60–92.

Tao, S. Y., Ni, Y. Q., Zhao, S. X., Wang, J. J. (2001). *Formation mechanism and forecast research for the rainstorm in China in 1998*, (Meteorological Press, China) (in Chinese)

Thompson, G., Rasmussen, R. M. and Manning, K. (2004). Explicit forecasts of winter precipitation using an improved bulk microphysics scheme. Part I: Description and sensitivity analysis, *Mon. Wea. Rev.*, 132, pp. 519–542.

Tory, K. J., Montgomery, M. T., Davidson, N. E. and Kepert, J. D. (2006). Prediction and diagnosis of tropical cyclone formation in an NWP system. Part III: Diagnosis of developing and nondeveloping storms, *J. Atmos. Sci.*, 64, pp. 3195–3213.

Trier, S., Skamarock, W., LeMone, M., Parsons, D. and Jorgensen, D. (1996). Structure and evolution of the 22 February 1993 TOGA COARE squall line: Numerical simulations, *J. Atmos. Sci.*, 53, pp. 2861–2886.

Vecchi, G. A. and Soden, B. J. (2007). Effect of remote sea surface temperature change on tropical cyclone potential intensity, *Nature*, 450, pp. 1066–1070.

Vecchi, G. A., Swanson, K. L. and Soden, B. J. (2008). Whither hurricane activity?, *Science*, 322, pp. 687–689.

Vigh, J. L. and Schubert, W. H. (2009). Rapid development of the tropical cyclone warm core, *Mon. Wea. Rev.*, 66, pp. 3335–3350.

Wang, B. and Li, X. (1992). The beta drift of three-dimensional vortices: A numerical study, *Mon. Wea. Rev.*, 120, pp. 579–593.

Wang, H., Wang, Y. and Xu, H. (2013). Improving simulation of a tropical cyclone using dynamical initialization and large-scale spectral nudging: A case study of Typhoon Megi (2010), *Acta Meteor. Sinica*, 27, pp. 455–475.

Wang, Y. (2002). An explicit simulation of tropical cyclones with a triply nested movable mesh primitive equations model-TCM3. Part II: Model refinements and sensitivity to cloud microphysics parameterization, *Mon. Wea. Rev.*, 130, pp. 3022–3036.

Wang, Y. (2007). A multiply nested, movable mesh, fully compressible, nonhydrostatic tropical cyclone model—TCM4: Model description and development of asymmetries without explicit asymmetric forcing, *Meteorol. Atmos. Phys.*, 97, pp. 93–116.

Wang, Y. (2008). Rapid filamentation zone in a numerically simulated tropical cyclone, *J. Atmos. Sci.*, 65, pp. 1158–1181.

Wang, Y. (2009). How do outer spiral rainbands affect tropical cyclone structure and intensity?, *J. Atmos. Sci.*, 66, pp. 1250–1273.

Wang, Y. and Holland, G. J. (1996). The beta drift of baroclinic vortices. Part II: Diabatic vortices, *J. Atmos. Sci.*, 53, pp. 3737–3756.

Wang, Y., Leung, L. R., Mcgregor, J. L., Lee, D.-K. and Wang, W. C. (2004a). Regional climate modeling: Progress, Challenges, and Prospects. *J. Meteor. Soc. Japan*, 82, pp. 1599–1628.

Wang, Y., Sen, O. L. and Wang, B. (2003). A highly resolved regional climate model (IPRC-RegCM) and its simulation of the 1998 severe precipitation event over China. Part I: Model description and verification of simulation, *J. Clim.*, 16, pp. 1721–1738.

Wang, Y. and Wang, B. (2001) *Present and Future of Modeling Global Environmental Change: Toward Integrated Modeling*, eds. Matsuno, T. and Kida, H., Chapter 2 "Toward a unified highly resolved regional climate modeling system", (Terra Press,Tokyo), pp. 29–48.

Wang, Y., Wang, Y. and Fudeyasu, H. (2009). The role of Typhoon Songda (2004) in producing distantly located heavy rainfall in Japan, *Mon. Wea. Rev.*, 137, pp. 3699–3716.

Wang, Y. and Wu, C.-C. (2004). Current understanding of tropical cyclone structure and intensity changes – a review, *Meteorol. Atmos. Phys.*, 87, pp. 257–278.

Wang, Y. and Xu, J. (2010). Energy production, frictional dissipation, and maximum intensity of a numerically simulated tropical cyclone, *J. Atmos. Sci.*, 67, pp. 97–116.

Wang, Z., Hu, J. and Ding, Y. H. (1991). The effects of flow patterns over the northwest pacific on typhoon tracks, *J. Applied Meteor. Sci.*, 4, pp. 362–368. (in Chinese)

Wang, Z. Q., Duan, A. M. and Wu, G. X. (2014). Impacts of boundary layer parameterization schemes and air-sea coupling on WRF simulation of the

East Asian summer monsoon, *Science China: Earth Sciences.*, 44, pp. 548–562.

Wang, Z. Z., Wu, G. X., Wu, T. W. and Yu, R. C. (2004b). Simulation of Asian monsoon seasonal variation with climate model R42L9/LASG, *Adv. Atmos. Sci.*, 21, pp. 879–889.

Warner, T. T., Peterson, R. A. and Treadon, R. E. (1997). A tutorial on lateral boundary conditions as a basic and potentially serious limitation to regional numerical weather prediction, *Bull. Am. Meteor. Soc.*, 78, pp. 2599–2617.

Webb, E. K. (1970). Profile relationships: The log-linear range, and extension to strong stability, *Quart. J. Roy. Meteorol. Soc.*, 96, pp. 67–90.

Wei, H., Fu, C. and Wang, W. (1998). The effect of lateral boundary treatment of regional climate model on the East Asian summer monsoon rainfall simulation, *Chinese J. Atmos. Sci.*, 5, pp. 779–790 (in Chinese).

Wei, X., Li, H. N. and Zhou, J. J. (2010). Analysis on the abnormal track before turning of typhoon "Sinlaku", *Marine Forecast.*, 27, pp. 35–38. (in Chinese)

Willoughby, H. E. (1988). The dynamics of the tropical cyclone core, *Aust. Met. Mag.*, 36, pp. 183–191.

Willoughby, H. E. (1990). Gradient balance in tropical cyclones, *J. Atmos. Sci.*, 47, pp. 265–274.

Willoughby, H. E. (2009). Diabatically induced secondary flows in tropical cyclone. Part II: Periodic forcing, *Mon. Wea. Rev.*, 137, pp. 822–835.

Wu, C.-C. (2001). Numerical simulation of Typhoon Gladys (1994) and its interaction with Taiwan terrain using the GFDL hurricane model, *Mon. Wea. Rev.*, 129, pp. 1533–1549.

Wu, C. C., Lee, C.- Y. and Lin, I.- I. (2007). The effect of the ocean eddy on tropical cyclone intensity, *J. Atmos. Sci.*, 64, pp. 3562–3578.

Wu, G. X., Chou, J. F., Liu, Y. M. and He, J. H. (2002). *Dynamics of the Formation and Variation of Subtropical Anticyclone.* (Science Press, Beijing) (in Chinese).

Wu, L., Wang, B. and Geng, S. (2005). Growing typhoon influence on East Asia, *Geophys. Res. Lett.*, 32, L18703.

Wu, L. and Wang, B. (2000). A potential vorticity tendency diagnostic approach for tropical cyclone motion, *Mon. Wea. Rev.*, 128, pp. 1899–1911.

Wyngaard, J. W. (2004). Toward numerical modeling in the "terra incognita"., *J. Atmos. Sci.*, 61, pp. 1816–1826.

Xu, J. and Wang, Y. (2010a). Sensitivity of the simulated tropical cyclone inner-core size to the initial vortex size, *Mon. Wea. Rev.*, 138, pp. 4135–4157.

Xu, J. and Wang, Y. (2010b) Sensitivity of tropical cyclone inner-core size and intensity to the radial distribution of surface entropy flux, *J. Atmos. Sci.*, 67, pp. 1831–1852.

Yang, B., Wang, Y. and Wang, B. (2007). The effect of internally generated inner-core asymmetries on tropical cyclone potential intensity, *J. Atmos. Sci.*, 64, pp. 1165–1188.

Yau, M. K., Liu, Y., Zhang, D.-L. and Chen, Y. (2004). A multiscale numerical chapter of Hurricane Andrew (1992). Part VI: Small-scale inner-core structures and wind streaks, *Mon. Wea. Rev.*, 132, pp. 1410–1433.

Yu, J. H., Tang, J. X. and Dai, Y. H. (2012). Analysis in errors and their causes of Chinese typhoon track operational forecasts, *Meteorol. Mon.*, 38, pp. 695–700. (in Chinese)

Yu, R., Li, W., Zhang, X., Liu, Y., Yu, Y., Liu, H. and Zhou, T. (2000). Climatic features related to eastern China summer rainfalls in the NCAR CCM3, *Adv. Atmos. Sci.*, 17, pp. 503–518.

Yu, X. and Lee, T.-Y. (2010). Role of convective parameterization in simulations of a convection band at grey-zone resolutions, *Tellus*, 62, pp. 617–632.

Yuter, S. E. and Houze, R. A. (1995). Three-dimensional kinematic and microphysical evolution of Florida cumulonimbus. Part II: Frequency distributions of vertical velocity, reflectivity, and differential reflectivity, *Mon. Wea. Rev.*, 123, pp. 1941–1963.

Zhang, B. H., Liu, S. H., Liu, H. P., et al. (2012). The effect of MYJ and YSU schemes on the simulation of boundary layer meteorological factors of WRF, *Chin. J. Geophys.*, 55, pp. 2239–2248. (in Chinese)

Zhang, D.-L., Liu, Y. and Yau, M. K. (2002). A multiscale numerical study of Hurricane Andrew (1992). Part V: Inner-core thermodynamics, *Mon. Wea. Rev.*, 130, pp. 2745–2763.

Zhang, G.-J. (1994). Effects of cumulus convection on the simulated monsoon circulation in a general circulation model, *Mon. Wea. Rev.*, 122, pp. 2022–2038.

Zhang, Q. H., Zhang, C. X. and Zhang, Z. F. (2007). Study on the uncertainty of ensemble forecasting of tropical cyclone, *Chin. J. Geophys.*, 50, pp. 701–706. (in Chinese)

Zhong, Z. (2006). A possible cause of a regional climate models' failure in simulating the East Asian summer monsoon, *Geophys. Res. Lett.*, 33, L24707.

Zhong, Z. and Hu, Y. (2007). Impacts of tropical cyclones on the regional climate: an East Asian summer monsoon case, *Atmos. Sci. Lett.*, 8, pp. 93–99.

Zhong, Z, Hu, Y. J., Min, J. Z. and Xu, H. (2007). Experiments on the spin-up time for the seasonal scale regional climate modelling, *Acta Meteor. Sin.*, 21, pp. 409–419.

Zhong, Z., Wang, X., and Min, J. (2010a). Testing the influence of WPSH on the precipitation over eastern China in summer using RegCM3, *Theor. Appl. Climatol.*, 100, pp. 67–78.

Zhong, Z., Wang, X., Lu, W., and Hu, Y. (2010b). Further study on the effect of buffer zone size on regional climate modeling, *Clim. Dyn.*, 35, pp. 1027–1038.

Zhong, Z. and Zhang, J. S. (2006). Explicit simulation on the track and intensity of tropical cyclone Winnie (1997), *J. Hydrodynamics*, 18, pp. 641–652.

Zhou, T. and Li, Z. (2002). Simulation of the east Asian summer monsoon by using a variable resolution atmospheric GCM, *Clim. Dyn.*, 19, pp. 167–180.

Zhou, T., Wu, B., and Wang, B. (2009). How well do atmospheric general circulation models capture the leading modes of the interannual variability of the Asian-Australian monsoon?, *J. Clim.*, 22, pp. 1159–1173.

Zhou, T. and Yu, R. (2005). Atmospheric water vapor transport associated with typical anomalous summer rainfall patterns in China, *J. Geophys. Res.*, 110, D08104.

Zhou, T., Yu, R., Li, H. and Wang, B. (2008). Ocean forcing to changes in global monsoon precipitation over the recent half-century, *J. Clim.*, 21, pp. 3833–3852.

Zhou, H., Zhu, W. J. and Peng, S. Q. (2013). The impacts of different microphysics schemes and boundary layer schemes on simulated track and intensity of super typhoon MEGI (1013), *J. Tropical Meteor.*, 28, pp. 803–812.

Zhu, Q. G., Lin, J. R. and Shou, S. W. (2000) *Synoptic Principles and Methods of Synoptic Meteorology* (Meteorological Press, Beijing).

Zhu, T. and Zhang, D.-L. (2006). Numerical simulation of Hurricane Bonnie (1998). Part II: Sensitivity to varying cloud microphysical processes, *J. Atmos. Sci.*, 63, pp. 109–126.

Zou, L. and Zhou, T. (2013). Can a regional ocean–atmosphere coupled model improve the simulation of the interannual variability of the western North Pacific summer monsoon?, *J. Climate*, 26, pp. 2353–2367.

Index

233